Applied Security Devices and Circuits

by P. Benton

Applied Security Devices and Circuits

by P. Benton

PROMPT® PUBLICATIONS

©2001 by Sams Technical Publishing

PROMPT® Publications is an imprint of Sams Technical Publishing, 5436 W. 78th St., Indianapolis, IN 46268.

All rights reserved. No part of this book shall be reproduced, stored in a retrieval system, or transmitted by any means, electronic, mechanical, photocopying, recording, or otherwise, without written permission from the publisher. No patent liability is assumed with respect to the use of the information contained herein. While every precaution has been taken in the preparation of this book, the author, the publisher or seller assumes no responsibility for errors or omissions. Neither is any liability assumed for damages resulting from the use of information contained herein.

International Standard Book Number: 0-7906-1247-X
Library of Congress Catalog Card Number: 2001093160

Acquisitions Editor: Deborah Abshier
Senior Editor: Kim Heusel
Editor: Jeannie Smith
Interior Design: Debbie Berman
Typesetting: Debbie Berman
Indexing: Kim Heusel
Proofreader: Kim Heusel, Lindsey Holloway
Cover Design: Christy Pierce
Graphics Conversion: Christy Pierce
Illustrations: Courtesy the author

Trademark Acknowledgments:
All product illustrations, product names and logos are trademarks of their respective manufacturers. All terms in this book that are known or suspected to be trademarks or services have been appropriately capitalized. PROMPT® Publications and Sams Technical Publishing cannot attest to the accuracy of this information. Use of an illustration, term or logo in this book should not be regarded as affecting the validity of any trademark or service mark.

PRINTED IN THE UNITED STATES OF AMERICA

9 8 7 6 5 4 3 2 1

Contents

1 General Security = Common Sense 1

Home Security 2
Vehicle security 3
Glossary of Security, Fire,
and Alarm Terminology 4

2 Basic Alarm Operating Principles 11

List of Input/Sensing Devices 12
    Wire loop 12
    Switches 12
    Light 14
    Sound 15
    Temperature 16
    Proximity 16
    Mains hum 17
    Pressure 17
    Radio 18
    Wireless systems 18
    Probes 19
    Cameras 19
List of Output Devices/Indicators 19
    Audible Devices 20
    Visual Devices 21
    Electromechanical Devices 21

Indicators and Alarms 22
    Simple Circuits 23
        Normally open switch circuits 26
        Normally closed switch circuits 26
        End-of-line resistors 32
    Two-Wire versus Four-Wire Systems 34
RC Timing Networks 37

3 Timing Circuits 37

Practical Considerations of Using
RC Timing Networks 38
Timing Circuit Using a Logic Gate 40
The 555 Timer Integrated Circuit 45

4 Circuit-Protecting Devices 51

5 Automobile Security 57

Automobile Circuits 58
    Automobile/Backup Battery Monitor Circuit 58
    Remote Key Fobs 61
    Lights-On Reminder 63
    Automobile Headlight Sensor 65
    Automobile Indicator Alarm 66
    Vehicle Knock Alarm 68
    Cigar Lighter Thief Confuser 69
    Automobile Headlight Delay 71
    Automobile (Daylight-Only) Reversing Alarm 72
    Automobile Battery Condition Monitor 74
    Automobile "Hot-Wire" Preventer 75

Thermostat/Ice/Temperature Alarm 78
Vehicle Tracking Devices 81
Automobile Infrared Tracker Beacon 86
Automobile Radio Theft Alarm 87
Automobile Hazard/Alarm Lamp Flasher 87

6   Operational Amplifier Circuits 91

Circuits Using Op-Amps 92
    Over-Temperature Alarm with Relay Output 92
    Frost/Ice Warning with Relay Output 93
    Dark Alarm with Relay Output 94
    Light Alarm with Relay Output 94
    Electronic Thermostat 94
    Touch/Hum Switch with Relay Output 95
    Battery Condition Monitor Using the 741 Op-Amp 97

7   Low to High–Voltage Circuits 99

Anti–Tamper Switch Protection 109

8   Miscellaneous Designs 109

Automatic Surveillance
Camera VCR Switching 112
Automatic Lighting Controlled
by Doorbell 114
Battery Backup Charging Circuit 117
Briefcase Anti-Theft Method 120
Combination Lock 121
Controlling Supply Voltage with Sensor Switches 125
Countdown (or Count-Up) Displays 129

Discrete Monitoring 132
   Telephone–to–Tape Recorder Interface 132
   Telephone Line In Use Indicator 133
   Telephone Line Tap Detection 134
   Telephone Transmitting Device 135
   Voice Transmitting Device 136
   Automatic Recording from Receiver 136
   Automatic Recording from Telephone 137
Door Intercom Unit 140
   "Dummy" Alarm Devices 143
Electromechanical Hit Sensors 146
Emergency Mains Failure Backup Lighting System 148
Fishbite Alarm 151
Floodlight Switching Alarm Indicator 152
Hum Sensor Switch, Through-Glass 155
Induction Transmitter and Receiver:
An "Electronic Key" 157
Infrared Transmitter Tester 160
Key Sentinel 162
Lie Detector/Damp Wall Detector 164
Loop Alarm System 166
Low-Cost Vehicle Immobilizer 168
Mains-On Alarm 169
Mains Supply Failure Alarm 172
Mains Supply Failure Alarm
with Latch 175
Metal Detector/"Hidden
Treasure Locator" 175
Multi-Camera Switching Controller 180
Nicad Battery Monitor 187
Night-Light Porch Switch 187
One-Second Beeper/Flasher 190
Refrigerator-Light Beeping Alarm 193
Snooper Scarers 194

Sound-Activated Switch 198
Sounder Circuits 200
Static Electricity/Lightning Detectors 204
Telephone Ringer Extender 207
Temperature-Activated
Appliance Switch 210
Timed Latch Alarm 212
Touch-Activated Alarm 213
Transformerless Mains Power Supply 216
Two-Transistor Multivibrator Circuit 217
Using LEDs as High-Power Indicators 220
Infrared "Night-Vision" Illuminator 222
Vibration Alarm Circuits 224
Video Transmission 228
Warning Beep Circuit 230
Water/Fluid-Level Detectors 231

Index 239

Introduction

This book is written specifically for the electronics technician who enjoys making projects that perform a particularly useful task, that of protecting the technician and his or her family, dwelling, and valuables. Electronics students also will find this book extremely helpful in generating ideas for school and college projects. This book not only contains information about circuits and household security but also includes a wide range of information on the more diverse applications available, such as ideas for indicators to monitor multiple situations; for example, vehicle tracking devices, dummy alarms, surveillance devices, and electric fence circuits, as well as many more designs not present in any one publication. An explanation of how the circuit works accompanies each circuit presented in the following pages, so that anyone who wishes to construct the circuit may gain knowledge in the application of electronic devices as well as acquire knowledge about them.

Because most of us work on a tight budget and complex circuits can cost a lot of money, this book presents no highly complex circuits. While readers should own or have access to basic electronic tools in order to test or experiment with these circuits, it is the author's intention that no expensive setting-up apparatus, such as oscilloscopes and frequency meters, be required. For the same reason, none of the circuits herein contain a memory device or controller necessitating use of software programming.

All components used in the following circuits are modern or standard and, at the time of this writing, easily available from many electronic component suppliers. Several circuits are "tolerant-friendly," letting you use alternative components from the ever-present electronics junk box.

IMPORTANT NOTICE

Applied Security Devices and Circuits contains ideas for circuits that can be built by anyone with basic understanding of electrical and electronic theory, enabling readers to build a comprehensive security system to protect themselves and their property, but please read the following carefully.

SAFETY WARNING, HAZARD

It is important to note that some of the designs in this book use high voltages that could prove dangerous or lethal if misused or handled incorrectly. No persons should attempt to build these designs unless they are fully competent and qualified to do so. If at all in doubt, obtain guidance from a competent person!

Fire Hazard

Many electronic circuits produce heat to some extent. Voltage regulators with insufficient heat sinking can quickly cause a burnt finger, as may voltage-dropping power supply resistors. Transformers also can become a fire hazard, as can high-current dry-cell batteries or accumulator batteries! Always use fuse protection and calculate power loss that is generated as heat.

Legal Warning

Some of the designs contained in this book make use of radio-frequency transmitter circuits. In many countries, it is illegal to own / manufacture / use transmitters without an appropriate government licence. Before contemplating building or experimenting with any transmitting device, you must check local, national, and international law; laws vary geographically, and severity of penalties can range from fines to imprisonment, or both!

Some designs in this book show connections being made to telephone circuits. The local, national, and international laws concerning connecting unauthorized equipment to telephone lines, or laws regarding interfering with telephone lines, also must be observed.

IF IN DOUBT, DO NOT DO IT!

Applied Security Devices and Circuits

by P. Benton

CHAPTER 1

General Security = Common Sense

When securing a premises, the first course of action is installing adequate locks, doors, and windows, yet with the best intentions in the world, it is easy to let down one's defenses. For example, I once was stirred from a heavy sleep by a knock on the door. Upon answering the knock, I found a freezing cold, elderly gentleman who introduced himself to me as a "double-glazing representative." I told the man that I would welcome a visit from an estimator regarding some work in need of doing and was about to give him my name and telephone number when my brain, up to then groggy from sleep, went into alert mode. What if the man was a burglar? Armed with my telephone number, it would be easy for him to check, at another time, whether or not anyone was at home. With this scenario in mind, I told the "representative" that I had no phone but would welcome an estimate by mail. Two years later, I'm still waiting for that estimate. (Maybe the double-glazing firm does business only with people who own telephones?)

The preceding anecdote brings us nicely to the following list of dos and don'ts, all of which just require common sense but still are useful reminders to us all.

HOME SECURITY

Always check that maintenance or utility personnel who call at your home have ID cards, and if necessary, telephone their offices to check them out. Do not always rely on ID cards or the telephone number printed on them. A boiler suit doth not make the gas man; do not be duped by someone standing at your door just because the person is wearing overalls and holding the mother of all monkey wrenches in his hand.

Try to have a panic-alarm button situated close at hand when answering the door, as well as one as near to the bed as possible, without running the risk of triggering it during a nightmare. Some systems employ a remote panic transmitter that can be carried in a pocket, hung around the neck, and so forth.

If it is not possible to visually see a caller at your door, take whatever steps are necessary to correct this situation, such as installing a peephole or closed-circuit TV system and adequate outdoor lighting, automated where possible.

Keep doors and windows secured if possible when in the building and always secured when the building is empty. Prevent heatstroke from harming pets by either fitting windows with locking facilities that allow the window to be slightly ajar, or leaving the air-conditioning on.

Test the security of a building by standing outside it, throwing your only key down an imaginary drainpipe, and then thinking of the easiest way to force an entry into the building. Now fix whatever means of entry you chose so you cannot force entry by that particular method. Repeat this procedure until you are satisfied that it would require a very savvy burglar to break in and steal your VCR. Next, repeat the procedure with your garage and any other outbuildings or with your car—whichever most likely would be next in line of attack by a frustrated burglar.

It does no good installing steel doors without also installing steel frames, and there is no use going to the lengths to install either if firefighters or paramedics must reach you in an emergency.

If you have committed the effort and cost of installing a security alarm system, use it all the time. The seven minutes it takes to drive your child to school every morning is probably the same seven minutes that a burglar has decided is the best time to break in and do whatever evil deeds he or she can do in five minutes.

Install an intercom so you do not have to open the door or speak to someone outside if you do not wish to do so.

Do not let a stranger enter your home to use the telephone, even if the person says it is an emergency or seems unwell. Lock your door and make the call yourself. A charge for the call, if applicable, is better than what may happen otherwise.

If you're going on vacation and have a security alarm installed in your home, local bylaws may require your notifying the police of the address of a key holder for use in the event that the alarm is triggered. Keep the number of people who know you're going away to a minimum. Ask a trustworthy neighbor to pick up your mail, newspapers, and so on (some burglary rings depend on inside information from newspaper delivery persons). Ask the neighbor to park a car in your driveway. Install randomly timed interior lights. Beware of people who show too much interest in your flight times or ask "Are all the family going?" type of questions. Have a friend or neighbor check your home every morning and evening for attempted forced entries.

Never give a stranger standing at your door any information that may be used against you, such as that you are alone. A simple door-to-door or telephone survey could tell a villain what material possessions may be found in your home, if the house is empty at certain times, and so on.

If using an answering machine, do not record a message saying that you are out; instead, record a less explicit message such as "Thank you for calling, I will return your call as soon as possible."

VEHICLE SECURITY

When on the road, if you see an accident, stay in your car and call 911 on your cell phone or from a public telephone in a safe area.

Avoid running out of gas in unsafe neighborhoods or on unlit roads by keeping a sufficient level of gas in your tank.

Try to park only in well-lit, patrolled, secure parking lots and garages. Always close your vehicle's windows and lock the doors.

Do not leave anything in view inside your vehicle; a thief might break a window for something as small as a (empty) pack of cigarettes. If the radio has a theft-protector allowing you to remove it from its place, lock it in the glove compartment or hide it under the seat before arriving at the parking lot, if possible.

When approaching your car, look around for anyone suspicious hanging around, and have the appropriate key ready to open the car door quickly. Look inside your vehicle before getting in to make sure no one is either lying in wait or lying down trying to hot-wire the ignition switch.

It is each individual's decision whether or not to lock the doors and close the windows when stopped at an intersection, as well as keeping valuables out of view. If your car is in an accident when your doors are locked and your windows closed, a rescuer would have greater difficulty in assisting you.

> **Note:** A security alarm system will not stop a burglar from breaking into a building. The main reason for an alarm system is to generate an alarm if the premises are being or have been attacked. It sometimes proves a sufficient deterrent simply to have a notice warning would-be intruders that an alarm is installed on the premises, so long as the pretense is kept up at all times!

GLOSSARY OF SECURITY, FIRE, AND ALARM TERMINOLOGY

Access code The code used to arm or disarm a security system.

Addressable device A fire alarm component that can be given an individual electronic and physical address or identifier.

Alarm signal A signal that indicates an alarm condition that requires immediate attention.

Alarm system A combination of input devices, a processing unit, and output devices.

Alphanumeric display A keypad that can display alphabetical letters as well as numbers; better than a row of LEDs (light-emitting diodes) when trying to remember which zone did what when an alarm is active.

Annunciator A unit that contains indicator lamps, displays, and so on, to provide status indication about a circuit, condition, etc.

Arm and disarm Turn the system on and off.

Away mode The armed mode of the system sensors are armed.

Backup battery A necessary part of a secure system; provides a rechargeable backup battery coming into play in the event of a mains power failure. These rechargeable batteries are maintenance-free but should be replaced every three to five years. A small, typically rechargeable, backup battery is included in many microprocessor-based control panels, so that the programmed memory and events recorder is maintained if mains power is cut or lost. An alarm bell box will also contain a battery so that the sounder will be activated for a period of time if the box or wiring to the box is attacked.

Base Main control panel of the security system.

Battery load test Sometimes part of a self-test procedure; places a load on the backup battery so as to simulate the condition of mains failure, and places a load on the battery that is similar to that of the circuit if in an alarm mode so as to ensure the battery is charged up correctly and will function as required.

Biometric Systems that include systems to measure voice patterns, voice recognitioning, handprints, and so on.

Bypass Temporary removal of a zone from operation so as not to trigger the system once it has been armed.

Central station A remote location that monitors signals from alarm systems. The supervisor of the system may then call the police and contact the owner of the attacked property, or other interested parties.

Chime Most systems allow a chime mode that gives a sound from the control panel if a door or window is opened; this acts as a low-level warning for a house if occupied, or for a small shop so as to alert the shopkeeper in a backroom that someone has entered the retail area, for example.

Closed loop A part of a system that is made up of components that are all normally closed switches and are all wired in series.

Communicator Optional part of a system that can be a dialer/modem, etc., to communicate information; for example, to a central station.

Current sensing See "voltage sensing."

Delay zones Some zones may be programmed to have a delay between triggering and activating the alarm; noticeably, the entry/exit point. Sometimes this may be so as to give burglars "enough rope to hang themselves" in a sting or surveillance operation.

Direct line Some high-level systems incorporate a dedicated line to the security services, with the dedicated line being entirely separate from the normal phone line. See "phone-line monitoring."

Disarm To switch off or deactivate the security system.

Door transmitter The remote component that monitors a connected sensor and sends a signal to a base if that sensor detects a violation.

Doppler sensor Ultrasonic Doppler sensors include a transmitter and receiver; a complex field is produced by the transmitter, which is read by the receiver as normal. If the field changes due to an object's entering this field, the sensor will then trigger the alarm.

Engineer's code The engineer will install an alarm and use the engineer's code to access code changes in order to test the system and then will install user-preferred settings, such as entry and exit times.

Entry delay A user/engineer- or factory-preset period of time that the user has to disarm the system after activating the system entry point.

Exit delay A preset period of time that the user has to leave the premises via the exit point after setting the alarm system.

Fire-alarm verification To prevent a full-blown alarm, some systems incorporate an "ignore the first trigger" facility, where the control panel resets the detector but activates the fire alarm if a second trigger occurs, perhaps after a predetermined period of time.

Fire zone An area covered by a smoke detector or heat sensor.

Force arming The act of arming a system even though one of the zones is troubled, perhaps due to a sensor malfunction or false alarm caused by cigarette smoke in an area where smoking is prohibited.

Hard-wired A system that uses wiring to connect the sensors to the control panel, as opposed to a wireless system that uses low-powered radio links.

Heat Detector A detector that will trigger an alarm if the predetermined temperature is reached. A rate-of-rise heat detector will be triggered if the device detects a temperature rise of a preset rate.

Home mode The armed mode of a system where only the door and window detectors are armed; that is, only the peripheral is armed, letting residents walk around the building without triggering the internal sensors.

Initiating device A component in the system that originates the transmission of a change of condition; for example, a smoke alarm.

Instant zone A zone that causes an instant alarm immediately after triggering.

Interior sensor A sensor that is apart from the peripheral sensors; for example, a movement detector or a contact on a bedroom door.

Keypad A keypad is used as a nonportable means of entering data into the base, or reading information from it, with the security disadvantage that if seen being entered, a code may be copied.

Local alarm A sounder; for example, a bell or a siren, that is activated at the residence, thus relying on scaring the intruder or upon neighbors acting by calling the police, etc.

Modem A modulator-demodulator that is used for transforming electronic information to and from another medium, such as radio or telephone wires.

Monitored alarm An alarm system that sends a signal to a central station to alert monitoring staff of system attack, monitoring conditions, and so on.

Motion sensor A device that detects movement within the covered area; may be ultrasonic, PIR (passive infrared), or microwave designs.

Multiple areas It sometimes is preferred to split up large zones, or an area covered by many detectors, into smaller areas that then may be armed/disarmed individually, perhaps giving an opportunity to exercise an authority-level system.

Open loop See "parallel circuit/loop."

Panic An alarm that is activated manually, even if a remote key fob is used, and may cause an audible alarm or silent panic alarm that is monitored by a security company.

Parallel circuit/loop A part of a system comprised solely of normally open switches or devices, which should be wired in parallel with one another.

Partition Zone within an area, which is programmed so as to be independent of whether or not the household is disarmed; for example, an unoccupied room harboring a gun collection may be programmed to be armed even though the rest of the household is disarmed.

Perimeter sensor The group of sensors that include all exterior door and window sensors.

Phone-line seizure when the system is activated, a communicator will seize the phone line and communicate information about the attack to the security company.

Phone-line monitoring Some systems use a communicator that can contact the security company automatically if an alarm system is activated. The system may send a signal every 24 hours or so to tell the company that all is well. If the phone line is damaged in some way, this condition may cause an alarm to be indicated within the protected building. Several large telecom companies will provide a service whereby, if a line of theirs is cut and this optional service is subscribed to, the telecom company alerts the security company if appropriate.

PIN Personal Identification Number.

Proximity In the case of a proximity sensor, if an object nears the sensor it triggers the sensor; no direct contact is required, and therefore, the range of the sensor may in theory be infinitely variable.

Remote control A device that allows arming and disarming of the system away from the actual base; for example, a remote keypad, a key-fob transmitter, and so on.

Remote host Where the main computer that controls a system is remotely located, allowing that computer to control many systems from its location.

Series circuit/loop See "closed loop."

Shock/seismic sensor Detects shock waves produced when glass, wood, or the like is damaged.

Stand alone A system where the unit functions as a self-contained device.

Switch sensing Hard-wired automobile system, which monitors changes caused by door switches and so on. House alarms use magnetic- or plunger-type switches.

24-hour zone A zone comprised of devices that are alert around the clock, such as detectors for fire, freezer failure, panic alarm, tamper alarm, and so on.

User code Once the system is installed by the engineer, the user is given instructions to enter his or her own code, which must be kept secret, not divulged, even to the engineer. This user code lets the user arm, disarm, and part-set the system in a fashion as shown by the engineer.

Violation A physical intrusion or an alert that a physical intrusion has taken place.

Voltage sensing Voltage-sensing and current-sensing devices work by sensing a voltage drop or increase in the current of a conductor. These systems can be used in automobile alarms, requiring very little wiring, but are not secure if a door switch is faulty/sticking, or if the interior courtesy lamp is burned out.

Walk test A facility that lets the user test the coverage of motion detectors without causing a full-blown alarm state; should be performed on a regular basis so as to test that devices are functioning properly.

Wireless system An installation that uses low-powered radio links between the sensors and the control panel, saving expensive wiring work; however, if the sensors are battery-powered, these batteries must be checked and replaced regularly.

Zone An individual, defined area covered by a sensor or sensors.

Zone bypass It is sometimes necessary to bypass or exclude a zone, perhaps in order for building maintenance to take place or for a bulky delivery to be made through a normally unused doorway.

CHAPTER 2
Basic Alarm Operating Principles

To provide a basic alarm system, just three building blocks are required:

1. The input or sensing device, which detects the presence of what we are monitoring for.

2. The signal processing block that processes the signal enough so as to activate the...

3. Output device or indicator.

A block diagram of this concept is shown in Figure 2.1.

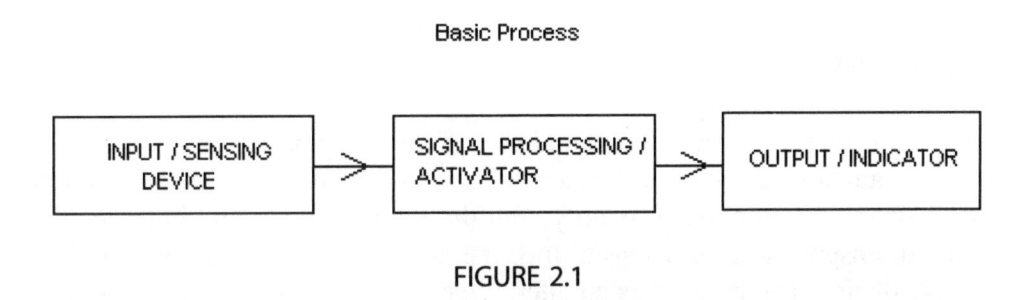

Basic Process

INPUT / SENSING DEVICE → SIGNAL PROCESSING / ACTIVATOR → OUTPUT / INDICATOR

FIGURE 2.1

LIST OF INPUT/SENSING DEVICES

A list of input/sensing devices could be endless, so following are just some of the common, and one or two of the more obscure, devices available.

Wire loop

The wire loop device is most commonly used for protecting "liftable" portable goods in shops and the like, activating the alarm when cut. This means that any attack requires renewal of the wire loop, or at the very least, unsightly repair. Wire loops are useful only if the goods to be within the zone have a handle for the loop to pass through.

These devices are used in the form of *window foil,* which adheres to a glass pane, so that when a window gets broken, in theory, so does the foil. If foil is used, it must be installed with care so as not to leave an area of unprotected glass large enough to be cut out and entered through by a pint-sized burglar. Overzealous varnishing when using a thick foil may cause the problem of a glass pane being attacked but held together by foil!

A wire loop may also be hidden inside a door in order to stop attackers from cutting or knocking out door panels so as to circumnavigate door/doorframe switches. Or this device may be included behind plasterboard that lines all walls or ceilings to guard against sledgehammer-wielding intruders. Wire loops may form a cheap alternative to vibration sensors or similar sensors.

Switches

Magnetic switches, found in the majority of installations, are in two parts: the magnet and a small, glass-encapsulated reed switch. When the magnet and the reed switch are in close proximity, and the latter is physically in the correct line (to ensure maximum sensitivity), the normally open contacts will close. These devices can be used as an alternative to "mechanical" tamper switches or opto devices, as safety switches to prevent machines from being powered up while open, for example. If a window has to be left open a few inches to allow ventilation on hot days, it is possible to fix another, second magnet in this position so as to maintain the loop.

Microswitches are small mechanical switches that, for our purposes, can be obtained with long activator arms. When positioned correctly, and with slight modification (bending with a pair of pliers), it is possible to make these switches so sensitive that they can be triggered by just a millimeter or so of movement on the activator arm. These switches are used mainly as "tamper" switches. They are fitted to the inside of an alarm control panel, bell box, remote sensor, or the like, so that if a box lid is opened, or if a box is physically torn from a wall, the alarm is activated. The switches usually are fitted so that they are continuously compressed and are released by the tampering.

Plunger switches are mounted inside holes that are drilled into the side of a door frame. When the door is closed, the spring-loaded plunger is depressed into the cavity, keeping the switch closed. When the door is opened, the plunger on the switch is then released, hence triggering the alarm. Note that with this type of switch, a second hole has to be drilled at a right angle to the main hole to allow wiring to be passed through.

Panic switches are manually activated switches that can activate your alarm system. These switches are usually large, visible, and robust, positioned next to a bed or by a front door but may be hidden, as with bank tellers' panic switches. These alarm triggers may cause bells to ring immediately or may operate a ***silent panic***, which is monitored by security services or may activate an automatic-dialing system to the police, for example. Some commercially available alarm systems require a special sequence of buttons on the control keypad to be pressed in the correct order so as to cause a silent panic alarm.

Trembler switches activate if the protected device is moved. The simplest trembler switch is a weight suspended on the end of a spring. Whenever the switch is knocked, the weight oscillates, swinging around and hitting a conductive ring that surrounds it, thus completing a circuit. These switches can be used in car alarms or, with a self-contained power supply (batteries), they may be hidden inside computers, video recorders, televisions, and so on.

Tilt switches contain mercury (now being replaced) or a ball bearing, which will roll onto a contact to complete the circuit. If the switch is not installed in the correct position, the switch may be somewhat insensitive or, even worse, activated either continuously or only when the thief is obliging enough to tip the item being stolen completely upside down. The tilt switch is often found in alarms that protect mobile or portable property that the owner wishes to keep immobilized, such as those items mentioned in the preceding paragraph.

Hall-effect switches are semiconductor-like devices that are affected by a magnetic field. They are used in some cellular telephones where the telephone handset contains the Hall-effect transistor, and the telephone holder contains the small magnet. These switches require that the magnet and detector be very close together.

Pressure mats are placed under carpeting and create a short circuit if activated. Smaller versions are available for fitting under stair carpets. Although relatively thin, pressure mats can leave a telltale bulge in an otherwise flat carpet that has been walked over several hundred times. To prevent a long-legged (or streetwise) villain from jumping over the one stair step with the pressure mat and missing it altogether, a suggestion is to conceal a pressure mat on two consecutive stair steps.

Light

The *light-dependent resistor,* or LDR, is the most common device used in amateur-type alarm designs; this is due to its flexibility, robustness, and relatively low cost. Various devices have different values, but in general, they have one thing in common: The resistance of the LDR cell is very high in darkness, typically two megohms, falling dramatically to typically 50 kilohms when fully illuminated. Visible or invisible laser lights can be used. Cheap laser pointer pens can provide an excellent basis for a visible-light system, using an array of mirrors to bounce the beam around; however, as with all lasers, these pens can cause temporary or permanent damage to the retina of the eye. Even if placed at knee height, they may cause suffering to wild animals and pets. Imagine waking up in the morning to find a dozen or so neighborhood cats stumbling blind in your backyard!

The *photovoltaic cell* is used less as a detector/sensor nowadays than in the past. It can produce a small voltage with a small current when illuminated so it is useful for constructing easy circuits. Small solar cells from scrapped calculators may be utilized.

Infrared and photodiodes and transistors are used a great deal as input sensors, although the infrared versions usually require an infrared source in this type of application. The *PIR,* or passive infrared detector, needs little introduction. The PIR works by sensing the heat that is radiated from a moving object. A system of lenses on the front of the detectors divides the field of view into separate zones, which are all focused on one central detector area within the body of the PIR. Because the PIR can detect motion best if the heat source cuts across the path of a zone, it is best to mount one of these devices so that the zone fan is walked across, rather than so that the body walks towards the unit. If a false alarm may be triggered by a static heat source, such as a radiator, place a strip of tape over that lens section to avoid false triggering.

Sound

Often overlooked, sound sensing has a large part to play in alarm design. A wide range of microphones and/or sound sensors are available to the designer. Electret microphones are cheap and widely available as electret microphone inserts. Although electret microphones do not produce a voltage—they need a supply usually derived from the main circuit—they are available in sizes down to 6 mm in diameter, with excellent reproduction and easy concealment. Used as mouthpieces in older telephones, carbon microphones operate as a varying resistance, as carbon granules are jiggled about by the sound; whereas dynamic microphones produce a small, varying voltage. Piezoelectric crystal microphones/detectors work upon the basis that, when a quartz crystal is compressed, it produces a minute voltage that can be amplified to a useful level.

Vibration sensors can work on any of the principles described in the preceding paragraph and can be utilized as a method for detecting glass breaking, doors or stud walls being vandalized or attacked, and so on. Glass-mounted sensors detect only the high frequencies caused when a pane of glass is shattered, and will not be falsely triggered by someone tapping the window or banging the glass with the palm of a hand. The other type of vibration detector/sensor is one that responds to bangs and knocks. With this type, it is up to the designer to choose a suitable sensitivity level and use an analyzer circuit that triggers the alarm only after a preset number of knocks.

Ultrasonic detection can be used in conjunction with an ultrasonic transmitter to cover a zone, using reflection or break-beam methods. An ultrasonic detector also can be used as a stand-alone device to detect bats.

Temperature

Many devices on the market remotely measure temperature or sense temperature changes. The most common of these devices is the thermistor, or "temperature-sensitive resistor," where the resistance changes with temperature.

Other temperature-sensitive devices available include capacitors and semiconductors, the latter of which can be obtained in the form of an integrated circuit designed for the sole purpose of temperature monitoring.

Proximity

Proximity sensors are available in an ever-continuing range. Inductive proximity sensors rely on the fact that if a foreign object gets close to an inductor, or coil, then the inductive properties of that coil are then altered, albeit slightly. The inductor may be alone, just being an inductance, or it may form part of a tuned circuit that either relies on the self-capacitance of the coil or functions in series or parallel with a capacitor. A typical example of this method is a metal detector or "treasure finder." When the search head (the coil) comes in close proximity to "treasure"—cigarette pack foil, aluminium ring-tabs, and such—the foreign metal alters the inductive properties of the coil, causing an imbalance in the system and thereby indicating a "find."

Magnetic sensing is based on the fact that if a magnet or a large, ferrous object is passed near a coil, then a small electric current is produced as the lines of the magnetic field cut across the coil. In the case of a ferrous object, the word *large* should be emphasized, because magnetic sensing relies on small magnetic areas in the object, or small fields produced as the object slices through the relatively weak magnetic field of the earth.

Mains hum

Unless in a geographical area completely devoid of mains electricity, there is a strong electromagnetic field caused by mains supply wiring. Almost everyone in the field of electronics has, at some time or other, been wiring an audio amplifier when he or she accidentally touched the input, or when a screened cable outer fell off, producing a loud, scary hum. It is possible to exploit this phenomenon by using it to detect people, produce touch-control designs, and so on. Another phenomenon that cannot be quantified is that of static electricity for detection purposes, but, however bizarre, it is another tool in the armory against intrusion.

Pressure

Pressure sensing is sometimes used to act as a monitor against intrusion in some commercially available systems for automobile protection, because if a window is broken or a door opened, the pressure inside the vehicle immediately changes. There are pressure sensors available that are based on a piezo-resistive principle, where a small current is applied to the device. Using a reference pressure, the output voltage rises proportionately with pressure increase, and vice versa.

Radio

Radio-frequency sensing methods can cover a range of diverse ideas. They are used in protective tabs and tags in stores and shops, with a small circuit being hidden under a store's ID label, for example, or inside a plastic box attached to the merchandise. These circuits may be a simple coil made from foil, with or without a diode, or may be a more complex circuit, such as a very low power transmitter. These circuits are "parasitic" insofar as they draw power from the transmitter positioned near a store's exits and checkouts. Various ingenious methods are employed here; for example, the load caused by the coil inside the merchandise label drains the checkout transmitters' power slightly, and thus can be measured by a circuit monitoring the supply current to the transmitter. The other method used in the preceding example could be that the low power transmitter inside the tag obtains power from the transmitter, in a parasitic fashion, in the checkout area. The small transmitter then retransmits a signal on a different frequency, which is then picked up to trigger the alarm.

Radio transmitters also may form part of a security system. Small, low-power transmitters can be installed in a few moments, to detect sounds or be activated by switches on gates or doors. They do not require wires to be fed back to the monitoring area, but they will require a power source—that is, batteries—that may need to be checked and refreshed at regular intervals, to maintain the system. One further problem with RF (radio frequency) links is that they can be monitored by an attacker with a scanning receiver and then perhaps jammed, or blocked out, by a copycat transmitting device.

Wireless systems

Most wireless systems usually work on the principle of the detector control panel continuously listening for a signal from the remote transmitters. This signal may be a "continuous" stream of data, which contains the address of each transmitter, with each transmitter working on a separate frequency, or, more commonly, the frequency is shared by every transmitter, where each one transmits a code every minute or so but gives a different code immediately when attacked. If the system waits for a "happy" signal from a transmitter but does not receive it within a timed period due to being flooded by a plain jamming signal, it may well go into alarm.

At the time of this writing, a novel device has been released on the market. It is a badge that is worn by the owner of a cellular telephone. When the ringer-off facility of the telephone is switched to silent mode, the badge will flash so as to silently alert the wearer that a call is being received. A similar circuit can be used as an alarm, to indicate that a nearby RF field has been switched on—such as a hidden transmitter bug!

Probes

Probes form a simple input method for sensing and monitoring the outside world. They can be used for monitoring water levels, humidity content, and any fluid that has even the smallest amount of conductivity.

Cameras

Cameras are becoming a larger part of security nowadays, because miniature CCDs (charge-coupled devices) are becoming cheaper. They are available in many shapes and sizes, including: monochromatic or color, with sound facilities, PIR activated, with video recorder auto switching, concealed inside clocks, looking through "pinholes" in wall partitions, worn on coat lapels, and so on.

LIST OF OUTPUT DEVICES/INDICATORS

Output devices may be broadly separated into three groups: audible devices, visual devices, and electromechanical devices.

Audible Devices

Buzzers, sirens, bells, bleepers, loudspeakers—the choice is up to the designer, who should choose the appropriate sounder for a specific purpose. A miniature siren that sounds impressive on the electronic workbench may be drowned out when positioned outdoors or in a large, noisy factory; that same sounder may split an eardrum or two if too loud in an enclosed space. Keep in mind that if the sounder is positioned as to be within easy reach of vandals, it must be enclosed in a suitable box, with appropriate anti-attack defense and tamper-proofing. Check with your local government to ensure that any external sounder you wish to install does not violate local ordinances with regards to maximum output in decibels (dB), time on or repeat cycle, or the type of noise allowable. An example of what may not be allowed likely would be a 150dB-police-sounding siren without a time-out.

Some sounders require only a supply voltage to make them operational, whereas other devices must be driven by a particular waveform. The latter therefore need a driver circuit and also possibly an audio amplifier circuit to give a sufficient power level to obtain high-level output. Despite these drawbacks, the idea of using this type of device should not be dismissed out of hand, because it is possible to design driver circuits that can produce startling, outrageous, or even science-fiction–like effects!

Electronic sounders are self-contained and include the driver circuit, which means that relatively speaking, they are "plug-in and go." Most suppliers will give the sound-output levels measured in dB, but be aware that not all measurements may have been taken at the same distance. The usual measured distance in specification data sheets is one meter. Sounders are available throughout the range, from little bleepers as found on keypads with negligible current consumption, up to and above 120dB ear-poppers with a current drain of two amps. Unless high-capacity batteries, mains supply, or accumulators were used, this power consumption would not be tolerable in a system powered by a humble torch cell.

It is worth noting that when designing a security circuit output, it is possible, if not a legal requirement, to have the sounder make different noises for different alarm situations, such as practice alarm, fire alarm, bomb alert, panic or intruder alarm, and so on. Some commercially available sounders have a row of switches, enabling the device to make tens of different noises either by your manually selecting the switch-bank settings or by your selecting them electronically.

Visual Devices

Visual indicators are used either to constantly give a readout of the information that has been processed, that is, a continuously variable state; or they are used as an indicator that an alarm has been activated or triggered, that is, the alarm is either on or off. Examples of the first type are voltmeters or ammeters, digital meters, LED or LCD displays with suitable driver circuits, video monitors, and so on. Examples of indicators for showing that an alarm has been triggered are filament lamps, strobes, LEDs, and the like.

Electromechanical Devices

The group of devices called *electromechanical devices* includes relays, solenoids, motors, pumps, and so on.

A relay is designed to be switched on by a relatively small current flowing through its coil, typically 100 mA (milliamperes) or so, which is provided by a transistor in the output/driver stage of a circuit. The coil becomes an electromagnet, pulling an activator arm, which closes (or opens) a pair of contacts. Depending on the current rating of these contacts, several amps at a high voltage may be switched, therefore it is possible to provide isolation between the low-voltage control circuit and mains power, etc.

Reed relays are small relays with a design similar to that of magnetic reed switch alarm contacts. Some types have a high resistance coil and may be driven directly from logic gates and buffers. Due to their small size, the maximum current and voltage ratings of the contacts are relatively small compared to their big brothers, but it is always possible to use the contacts of a small relay to power-up the coil of a much larger relay so as to provide heavy current switching.

Solid-state relays are not mechanical in nature but contain a circuit, typically including opto-couplers, thyristors, and the like. These devices are silent in operation, have no moving parts to wear out, and provide good isolation between input and output. The major drawbacks with solid-state relays are the extra expense; availability in the case of replacement; and, unlike transparently cased relays, solid-state relays do not give you the satisfaction of seeing or hearing them click in and out, a bonus when faultfinding!

> **Note:** Mechanical relays do have at least one drawback: If a circuit containing a relay is knocked hard enough, or receives similar rough handling, the contacts will bounce and momentarily touch or separate. If the circuit has a self-latching design, an alarm state will occur every time the device receives a good clout, providing a source of endless amusement for local idiots.

Solenoids are constructed from a coil of wire that forms a tube. Inside the tube is a rod of ferrous material, with a spring bias. When the coil is activated, it then tries to pull the rod toward it since it now has become a powerful electromagnet. The rod has an extension that sticks out of one end of the coil and is used for locking or unlocking doors, holding doors open, closing fire doors on command during an alarm, and so on.

Electrical motors and pumps usually have a large current consumption, especially when starting up or under a load. Because the load is constantly variable, and start-up current may look almost like a short circuit to the supply device, it is recommended that whenever possible these devices be driven by a relay and not a transistor. If a solid-state device is used to drive motors—or any other high-current device, such as high-wattage lamps or bells—then a semiconductor of sufficient rating and adequate heatsinking must be provided to protect the semiconductor from thermal damage.

INDICATORS AND ALARMS

This section shows how a basic alarm system can be designed, using either normally open or normally closed switches, and the inherent security for both switch types when used in two-wire and four-wire designs.

Simple Circuits

As seen in Figure 2.1, an alarm has three general sections, including a signal processor. These few introductory circuits, however, have only two components: a switch for the input device and a lamp, or buzzer, for the output device (see Figure 2.2 and Figure 2.3). In some applications, these basic circuits may suffice in providing adequate indication, such as a simple system that allows a shopkeeper in a backroom workshop to be aware that someone has entered the shop from the street. The drawback with a simple on or off circuit is that the output device is on only when the switch is on, or off when the switch is off. If it is required that the output device be triggered and stay on for a predetermined time, then a timing circuit would be employed, whereas if it is required that the output stay on until the shopkeeper resets it, then a latching circuit is used.

A simple example of processing a signal is shown in Figure 2.4 and Figure 2.5. In this application, it is necessary to measure a relatively high resistance: that of human skin. If the circuit shown in Figure 2.4 was used to attempt to light the lamp, then the lamp would illuminate only if the two probes were put

Simplest form of indicator using a normally open switch

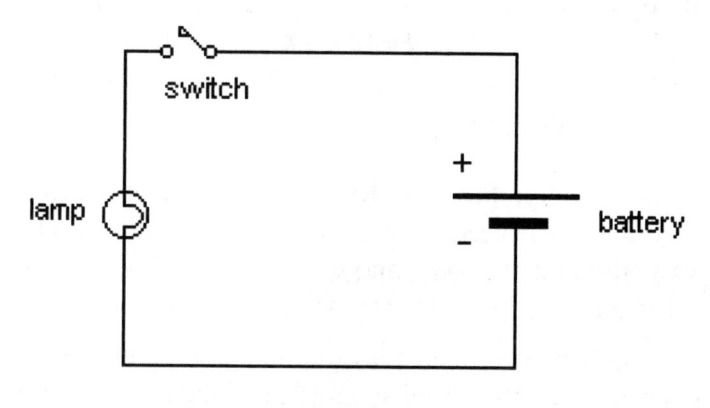

FIGURE 2.2

FIGURE 2.3

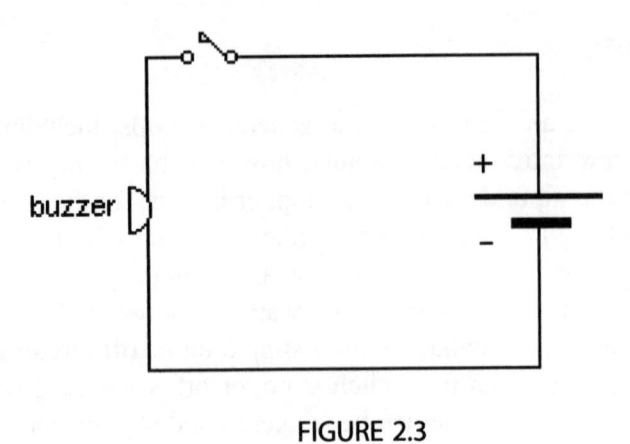

Single transistor amplifier

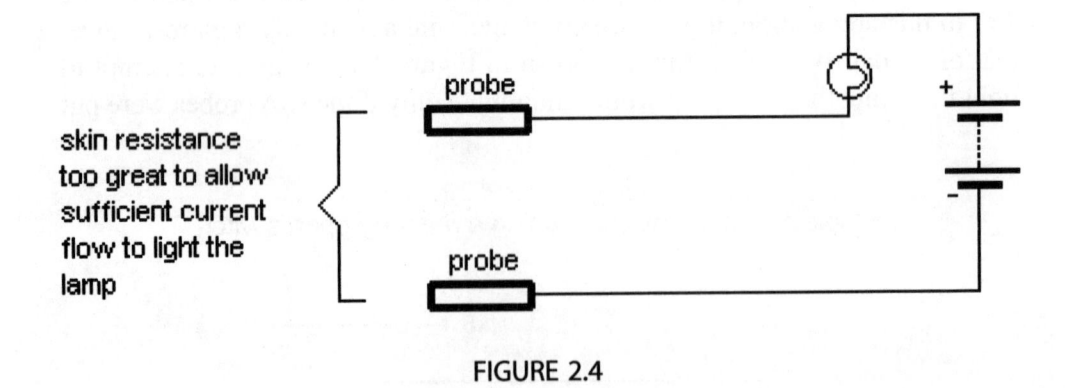

skin resistance too great to allow sufficient current flow to light the lamp

probe

probe

FIGURE 2.4

together, but not if the probes were held in the hands. However, if the circuit shown in Figure 2.5 was used, then the transistor would "amplify" the small current passing through the body, and the lamp would be illuminated; that is, the transistor is the processing section, the probes are the input, and the lamp is the output device.

BASIC ALARM OPERATING PRINCIPLES 25

transistor behaves as a "current amplifier", lamp will now light

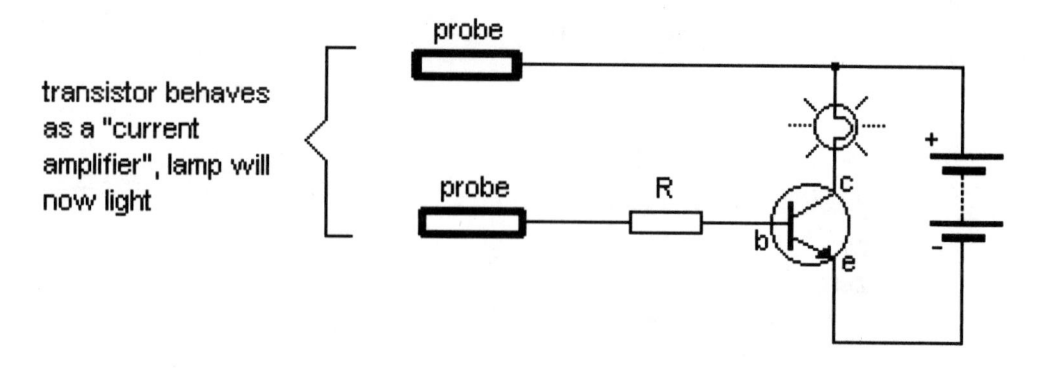

FIGURE 2.5

Simple lie detector

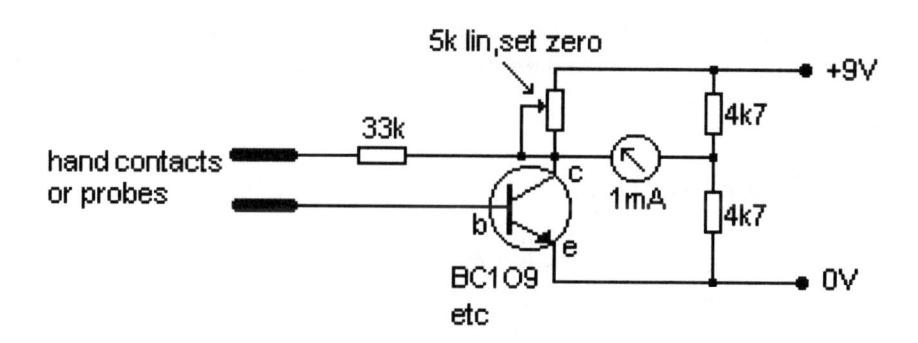

FIGURE 2.6

A simple "lie detector" is shown in Figure 2.6, where a meter indicates the moistness of the hands, related to the resistance of the "suspect." The transistor acts as a variable leg in a bridge circuit that consists of the transistor and the three resistors. This circuit demonstrates the use of a bridge circuit (see Figure 2.7), which is used in many measurement and sensor circuits. The variable leg, or element, may not be a transistor but may be many devices, such as a thermistor, LDR, or the like. Although a meter was used in the example of the

Bridge circuit

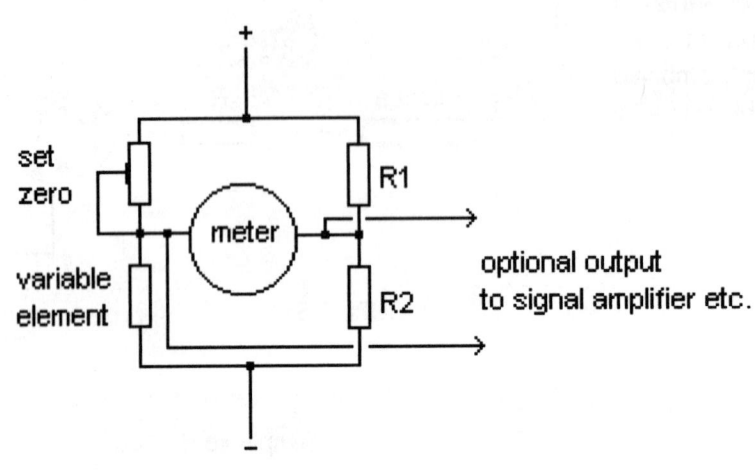

FIGURE 2.7

lie detector circuit, the optional output connections may go to the inputs of a signal amplifier such as an operational amplifier (that is, an op-amp).

Normally open switch circuits

In Figure 2.2, we used a normally open switch to turn on the output device. A practical example of the normally open switch would be a pressure mat, which is constructed from two conductive foil plates separated by a nonconducting foam cell. The method of connecting a pressure mat is shown in Figure 2.8, which shows how the circuit is completed whenever the mat is trod on.

Normally open switch "loops" have all their components wired together in parallel, as shown in Figure 2.9. This is so that if any of the switches—whether the panic button, switch, or pressure mat in the "loop"—are activated, that is, closed, then this circuit becomes short-circuited, thereby triggering the alarm.

BASIC ALARM OPERATING PRINCIPLES 27

Normally Open (n.o.) Contacts

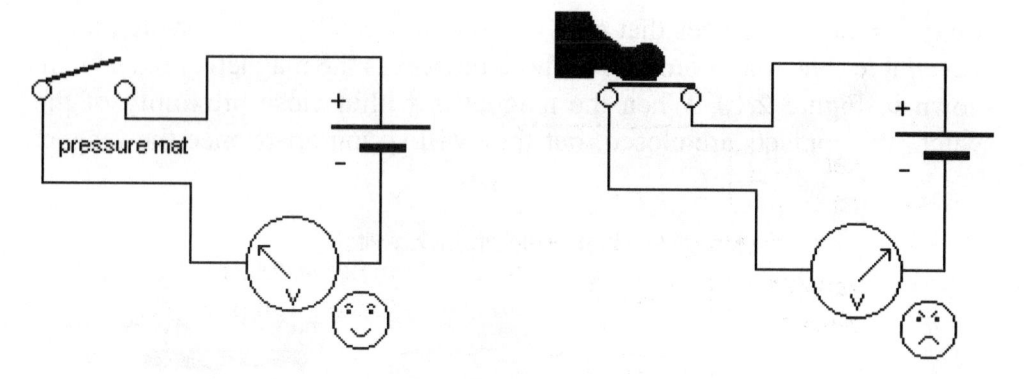

A normally open contact stays open until activated

FIGURE 2.8

Example of a normally open loop,
devices wired in parallel

panic button n.o.

switch

common n.o. n.c.

pressure mat

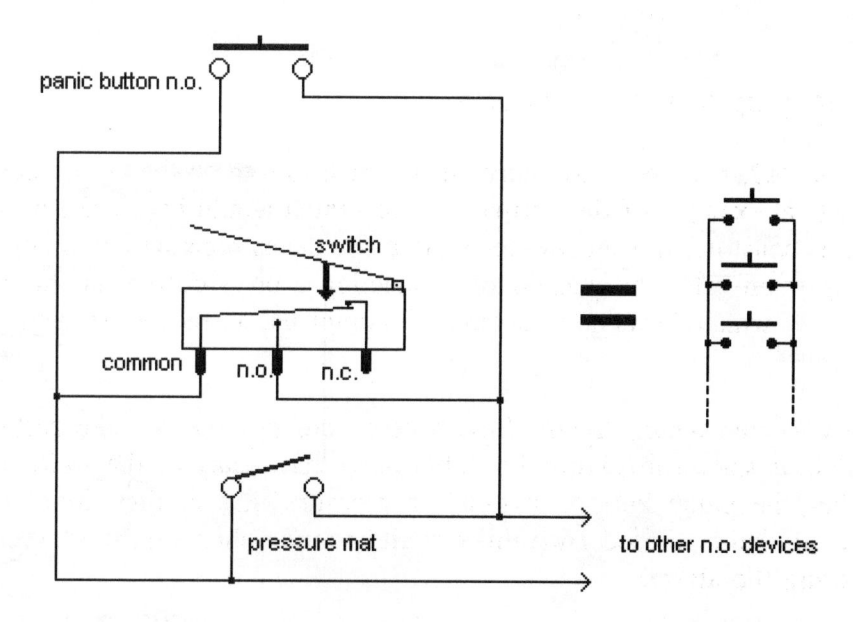

to other n.o. devices

FIGURE 2.9

Normally closed switch circuits

A large choice of devices that may be used in normally closed switch loops is available. The most common of these devices is the magnetic reed switch, shown in Figure 2.10. When the magnet is within close proximity of the switch, the contacts are closed, but they will spring apart once the magnet

Magnetic door and window switch

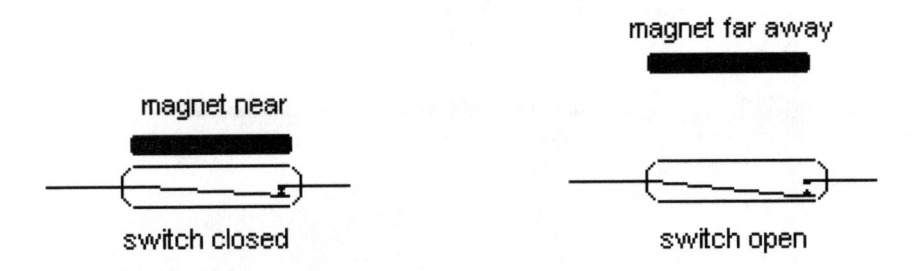

magnet near — switch closed

magnet far away — switch open

FIGURE 2.10

Magnetic door and window switch

magnet in close proximity, aperture closed — current flows

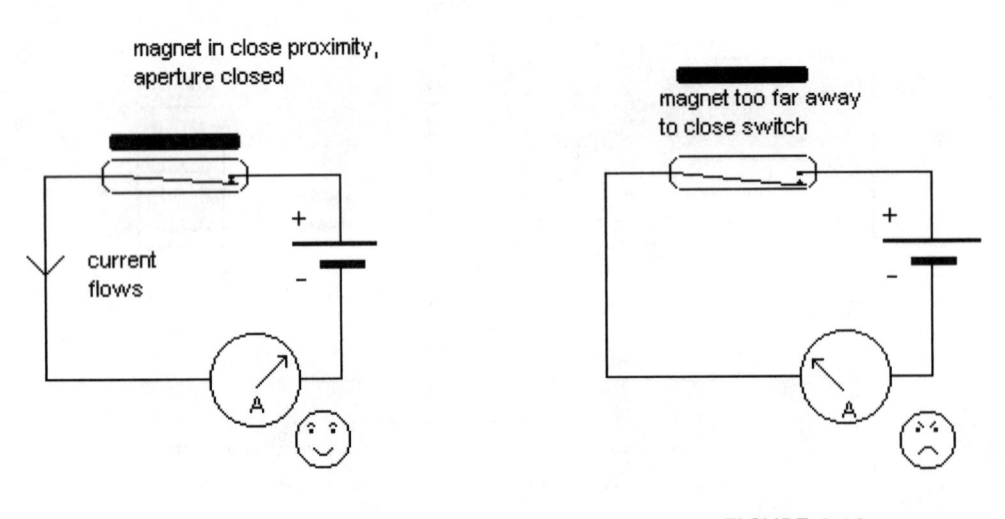

magnet too far away to close switch

FIGURE 2.11

FIGURE 2.12

moves away. Figure 2.11 shows that while the switch contacts are closed together, current flows through the circuit, but when the magnet is moved away, as shown in Figure 2.12, the current no longer flows because the circuit is now open.

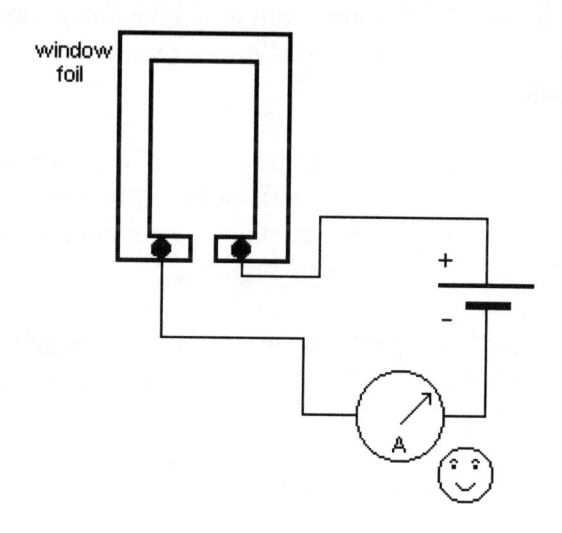

FIGURE 2.13

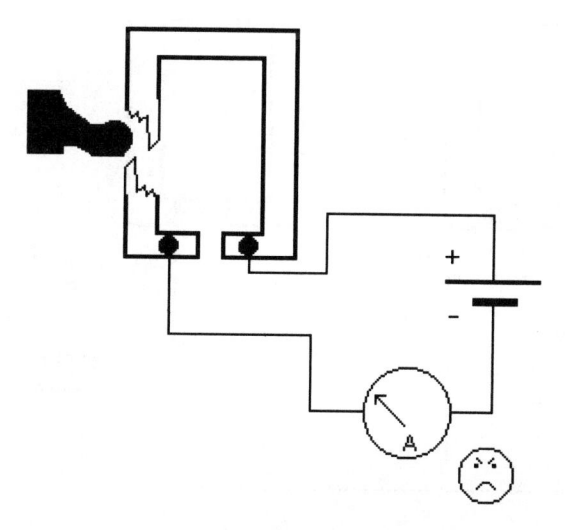

FIGURE 2.14

Another normally closed device that still is common is the window foil loop. This device acts as a closed circuit, allowing current to flow, as seen in Figure 2.13, but becomes an open circuit when attacked, as shown in Figure 2.14. When window foil is installed, it is very important that any area large enough for a person to crawl through is not left uncovered, and that lightweight goods such as jewelry cannot be grabbed, through any small hole that may be carefully cut into the glass, and reached either by a hand or by means of an extending-arm device.

All normally closed devices in the loop are wired together in series, that is, end to end, so that if any switch is activated—forced open—then the loop becomes open-circuit, so triggering the alarm. An example of this method of wiring up several devices is shown in Figure 2.15.

Daisy-chaining is a term sometimes used to describe the series connection of a number of normally closed switches that form a loop. In practical applica-

Example of a normally closed loop, devices wired in series

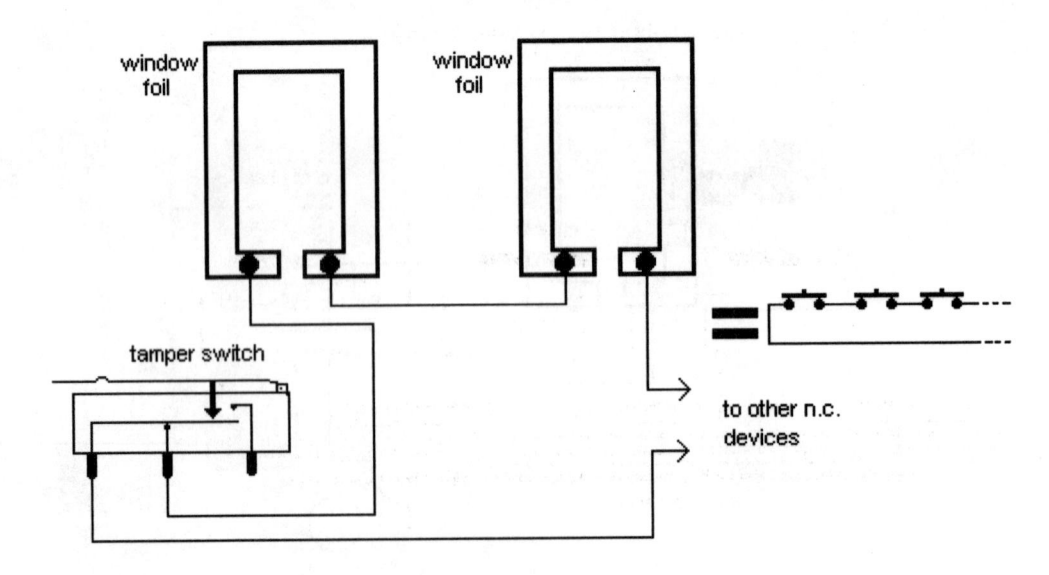

FIGURE 2.15

tions, a further wire or pair of wires are used within the same cable network. This wire or pair of wires are used so as to provide a normally closed, 24-hour tamper protection warning from the cable being cut, either by accident or design. Imagine an installation in a building that has all the protection circuits disarmed during normal "open" hours. Someone may think of cutting the alarm cable so that when the system is armed at the end of the business day, the alarm will not work, either because they might think that a severed cable will stop the alarm system from working, or more correctly, that the owner will find it impossible to set the alarm unless an engineer has been called to the business location, therefore causing remedial work to wait until the following day. If this 24-hour wiring is utilized, cutting this cable will cause an instant alarm, alerting the owner to take immediate action and scaring the culprit into thinking twice about following his or her plan. Figure 2.16 shows a typical daisy chain of magnetic door and window switches, whose connections are terminated at points A and B. Points C and D are connected in series to other tamper devices, or in series with other 24-hour devices if the circuit design cannot accommodate the extra input. Note that wires C and D are end-terminated together at the end of the daisy chain.

Wiring diagram of door and window contacts configured as a "daisy-chain"

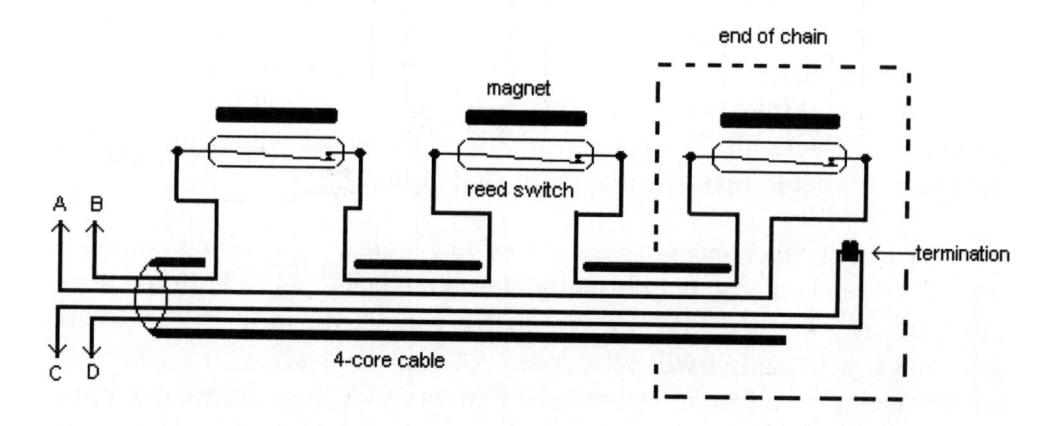

FIGURE 2.16

Other 24-hour circuits should include fire alarms, smoke alarms, gas alarms, flood alarms, freezer alarms, and so on. All important sensors should be of the normally closed variety—imagine a fire sensor loop that was normally open, on just a two-wire system, when the fire could burn through the cable! If it is at all possible, it may be prudent to use separate zone inputs for different loops and devices.

End-of-line resistors

Until now, we have looked at switching in only one or the other of two states: ones that are normally open—that is, short circuits when attacked—or normally closed—that is, open circuits when attacked. For each case it is necessary to have a separate design of circuit to be able to act on the two different states. As you can see in Figure 2.17a and Figure 2.17b, with an additional resistor, it is possible to poll both types of loop with a circuit that can recognize three states: open circuit, resistive circuit, or short circuit. However, probably the best use of this concept is that shown in Figure 2.17c, which allows both normally open and normally closed contacts to be included in the same circuit. With this circuit, if a normally closed switch becomes open, the end terminals see an open circuit; if a normally open switch becomes short, then a short circuit is formed. As with the previous two examples, the circuit will look like a 10k resistance if not attacked. It is not necessary to connect the resistor at the end of the line as drawn. With the first two examples, it is possible with ease to physically connect the EOL (end-of-line) resistor at the control panel; indeed, it is possible to do so with the third example as well, with a bit more fiddling, which is a positive boon with all the space available at the panel when compared to the space available inside a magnetic switch body!

Another use of this concept is illustrated in Figure 2.17d, which combines normally open contacts and normally closed contacts, plus additional resistors that are switched into the circuit by various devices. If one of the detectors is activated, it will introduce a 1k resistance across the 10k resistor, producing an effective resistance of around 910 ohms across the terminals of the circuit. Once again, it is only necessary to feed this information

BASIC ALARM OPERATING PRINCIPLES   33

into a circuit that can discriminate between these readings so as to produce a design that requires only one pair of terminals for a complex, hybrid loop of detectors. The values of resistance mentioned here, 10k and 1k, are purely

Example of the use of EOL-end of line resistors

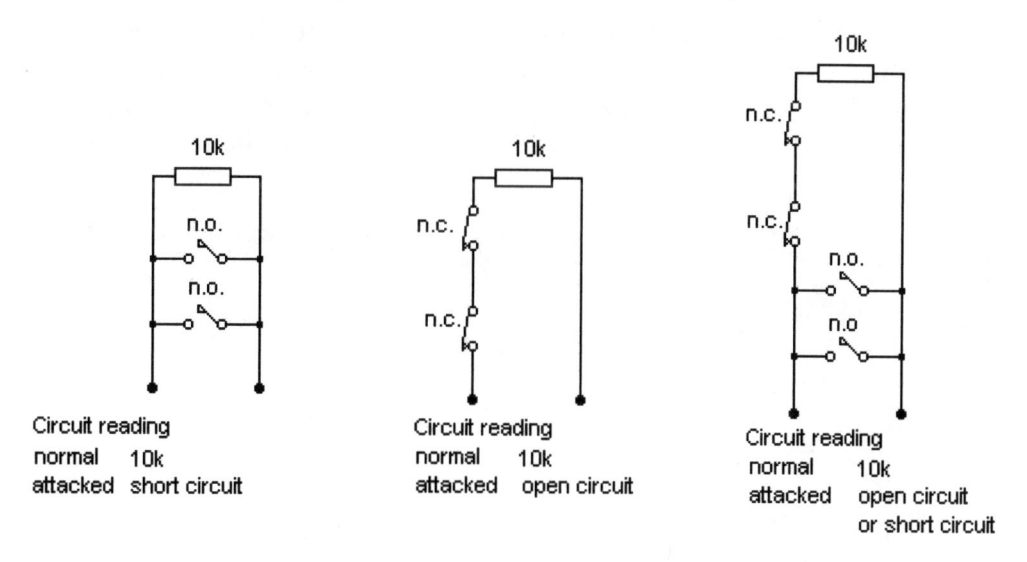

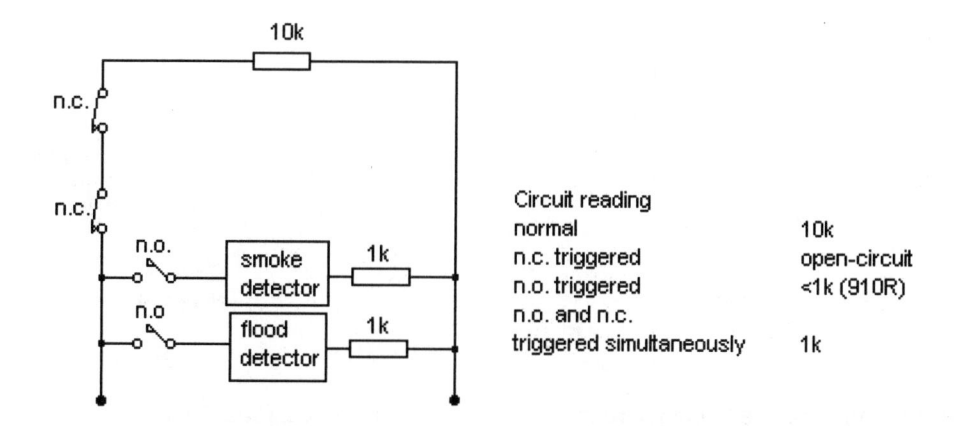

Circuit reading
normal   10k
attacked   short circuit

Circuit reading
normal   10k
attacked   open circuit

Circuit reading
normal   10k
attacked   open circuit
           or short circuit

Circuit reading
normal   10k
n.c. triggered   open-circuit
n.o. triggered   <1k (910R)
n.o. and n.c.
triggered simultaneously   1k

FIGURE 2.17d

theoretical and may be completely different values, but please note that some commercial devices available may support only their own values.

Two-Wire versus Four-Wire Systems

As seen in previous examples of simple circuits, it is possible to have a loop system that utilizes only two wires to connect the switches to the control panel or processor. Unfortunately, using normally open switches on a two-wire system presents a very poor level of security, as can be seen in Figure

Problem of using a two-wire system with normally open contacts

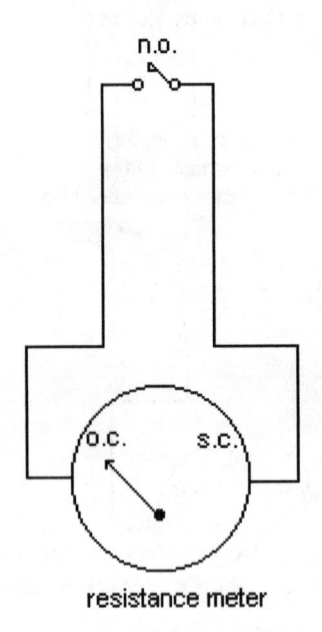

resistance meter

resistance meter

monitoring circuit sees open circuit under condition

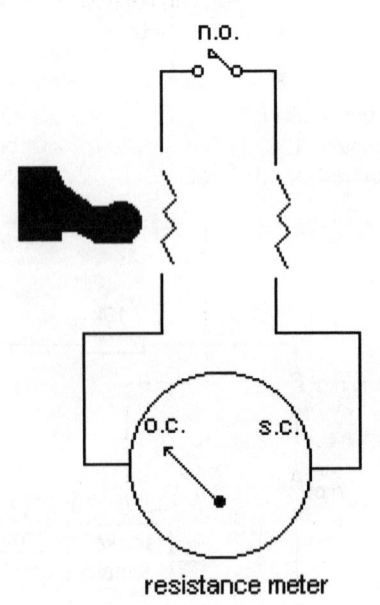

monitoring circuit still sees open circuit although attacked

FIGURE 2.18a

FIGURE 2.18b

2.18a and Figure 2.18b. The end terminals of a circuit containing only normally open switches will recognize the circuit as being open-circuit until the switch becomes closed, hence initiating an alarm status; however, if the cable connecting the switches to the processor becomes damaged and thus open-circuit, the panel will still think that the loop is satisfactory, even if the switches are activated after the cable has been damaged. Note that, in the illustrations, "o.c." denotes open-circuit, "s.c." denotes short-circuit, "n.o." denotes normally open contact or switch, and "n.c." denotes normally closed contact or switch.

A two-wire loop system containing normally closed switches will constantly be recognized as a short circuit until the switches are activated, which will

Good performance using a two-wire system with normally closed contacts

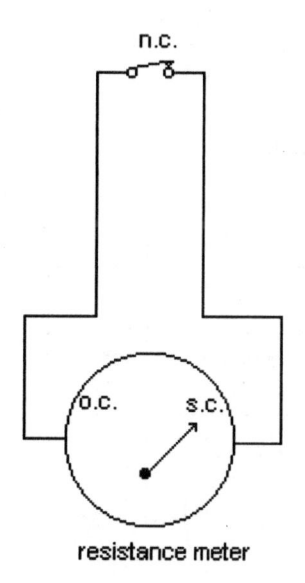

monitoring circuit sees short circuit under condition

FIGURE 2.19a

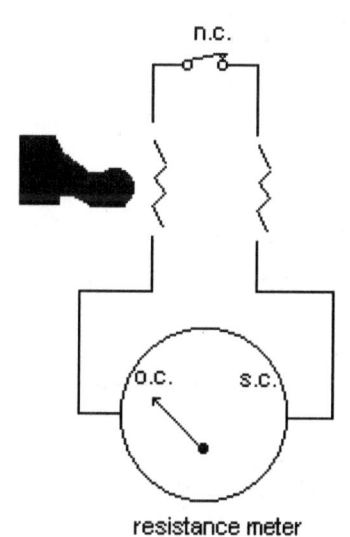

monitoring circuit now sees open circuit when attacked

FIGURE 2.19b

then produce an open circuit so as to initiate an alarm. This normally closed loop has the advantage that if the cable is cut, as shown in Figure 2.19a and Figure 2.19b, then this cut will also represent an open circuit and therefore an alarm will be triggered.

If a four-wire system is implemented, then the two extra wires may be used as a closed loop if they are terminated at the end of the system, as shown in Figure 2.20a and Figure 2.20b, to watch over a system of normally open switches. This pair of wires also may be used as extra protection for a normally closed loop, as shown earlier in the example of the "daisy-chaining" of normally closed switches.

Good performance using a four-wire system with normally open contacts and an anti-tamper pair

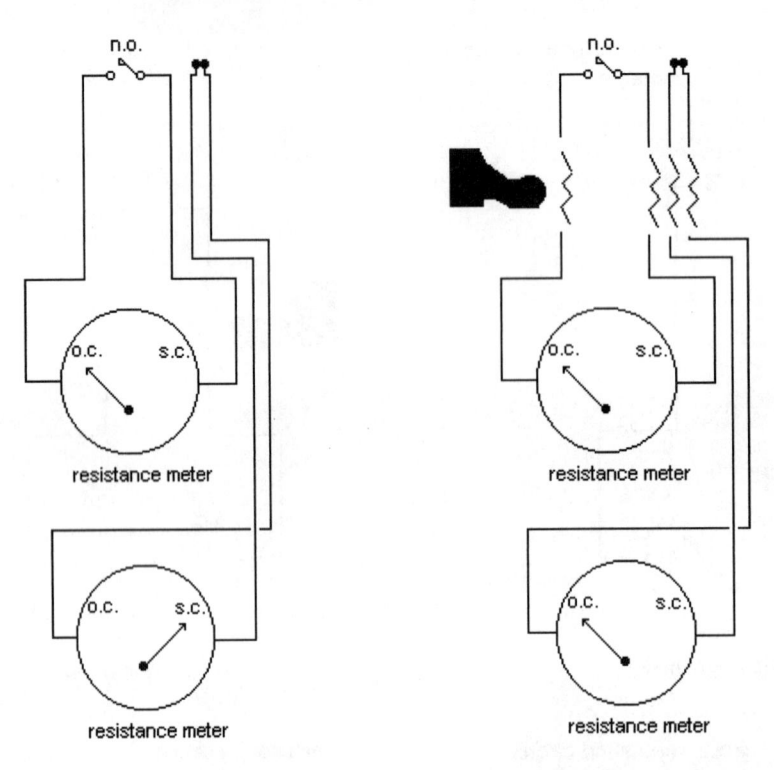

monitoring circuit sees open circuit when attacked

FIGURE 2.20a and b Monitoring circuit sees short-circuit when system is not compromised.

CHAPTER 3

Timing Circuits

Timing circuits perform the task of providing an output for a predetermined period once the circuit has been triggered in some way or another. The output period may be either high or low, depending on the design, and may be controlled by an RC (resistance and capacitance) combination. Other timing circuits will produce an output whose state may continuously alter from high to low, such as those used in siren outputs and the like.

RC TIMING NETWORKS

This useful network contains nothing more than a resistor and capacitor in series with each other, as illustrated in Figure 3.1. The graph in Figure 3.1 shows how the voltage measured across the capacitor builds up to (nearly) the value of the supply voltage. A quick calculation, multiplying the value of the resistor R in ohms and the value of the capacitor C in farads gives a result of the time taken, in seconds, for when the curve reaches approximately two-thirds of the supply voltage.

For example: R = 100k, C = 100uF, and V supply = 12V

(100,000 ∞ 100) / 1,000,000 = 10

Basic resistor and capacitor timing circuit

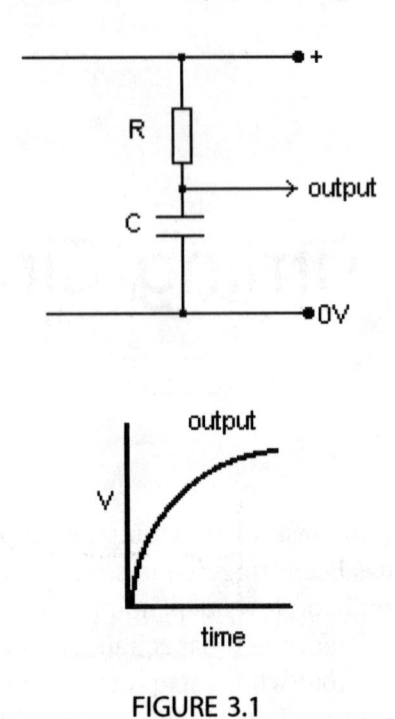

FIGURE 3.1

That is, 10 seconds to reach 2/3rds of 12V (approximately 7 volts)

Note that in calculations, the unit of capacitance, the farad, is used, whereas the capacitor in practice is available in the much smaller unit, the microfarad, which is one millionth of a farad. This anomaly means that if an unsuspecting novice designer works on a circuit calculation without doing the arithmetic correctly, the end results will be somewhat inaccurate.

Practical Considerations of Using RC Timing Networks

Refer to Figure 3.2a, showing an RC network connected to the base of a transistor, in order to turn on an LED after a predetermined time. When the

Simple time delay for LED

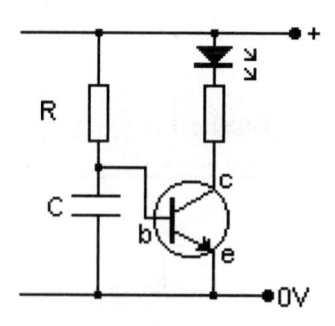

FIGURE 3.2a

base of transistor

0.7V

time

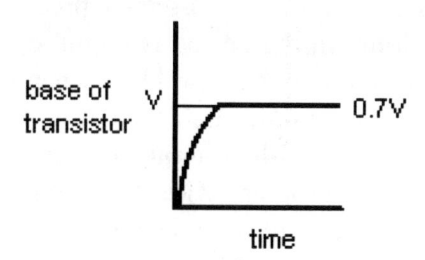

FIGURE 3.2b

circuit is powered up, the capacitor begins to charge up toward the supply voltage. The maximum voltage that can appear between the base and emitter of a bipolar transistor is around 0.7V, so as soon as the voltage on the base reaches this value, the capacitor ceases charging up, which means that the base voltage will remain the same (see Figure 3.2b). Because the transistor is now switched on, the LED will be illuminated. Note that the switching waveform will not be very clean, or sharp, because of the switching characteristics of the transistor, so if you carefully observe, you will see that the LED becomes illuminated slowly at first before becoming fully lit. The RC timing period (see Figure 3.2c) will be slightly longer than calculated because there will be a slight leakage through the base-emitter junction of the transistor, which means that, although negligible, the capacitor will take longer to charge up toward the switch-on voltage.

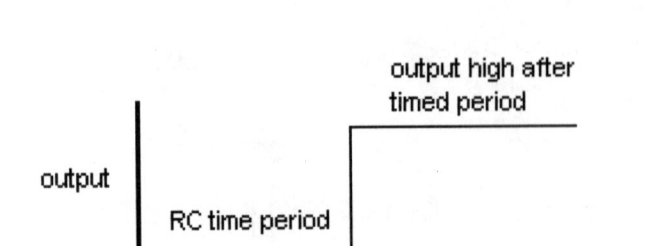

FIGURE 3.2c

In addition to the problem of slow switch-on, the preceding circuit has another inherent problem. If a long timing period is required, then the capacitor has to have a large value, which may be a problem when it comes to expense or physical size of the constructed project. Also, a large amount of the "potential capacitance" is wasted, because the circuit turns on when the base voltage reaches only 0.7V. Another problem with this circuit is that if the value of timing resistor R has to be large to provide a long time delay, then there may not be sufficient base current to enable the transistor to turn on fully, if at all.

Timing Circuit Using a Logic Gate

Figure 3.3a shows an RC timing circuit using a logic gate to replace the bipolar transistor used in the previous example. The gate could be a NAND gate from a 4011 integrated circuit in this particular application, with the inputs linked together; this means that the output will be inverted, as shown in Figure 3.3c. The advantage of this circuit can be clearly seen by referring to Figure 3.3b, which shows that the timing period will not end until the input of the logic gate, point A, reaches around one half to two-thirds of the supply voltage, depending on the family of integrated circuit being implemented in the design.

It may not be acceptable for the output of the timing circuit to be upside down, in which case the circuit shown in Figure 3.4a can be used. A second NAND

Timing circuit using a logic gate

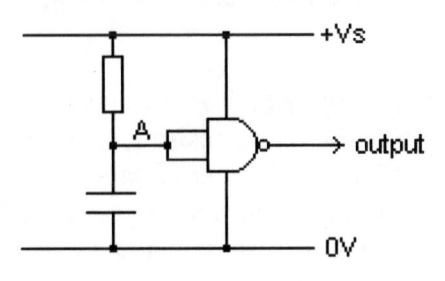

FIGURE 3.3a

voltage at point A   1/2 Vs, 2/3Vs etc.

time

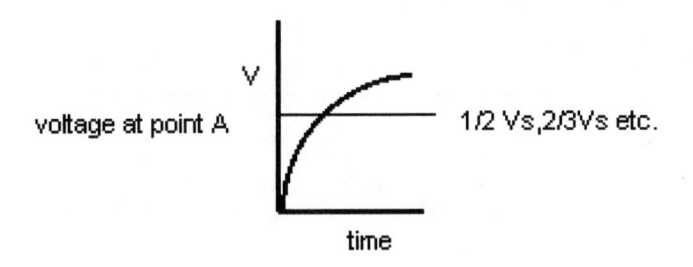

FIGURE 3.3b

output high for timed period

output

time

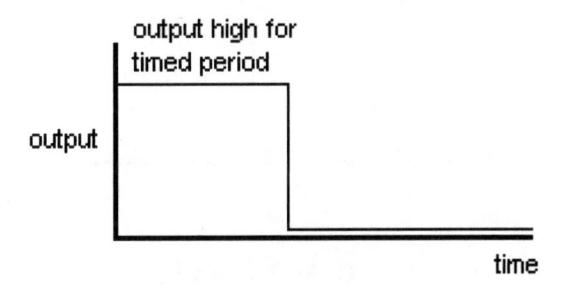

FIGURE 3.3c

gate follows the original gate, and because both inputs of this second gate are connected, the output is inverted to produce an output waveform as shown in Figure 3.4b.

When considering the relatively high impedance of a logic gate as compared with the impedance of the bipolar transistor, combined with the fact that the gate will not trigger until a much higher voltage is reached, we can appreciate that the logic gate has an important role to play in RC timing circuits. Much longer time delays are possible, and that valuable capacitance mentioned earlier is not "wasted."

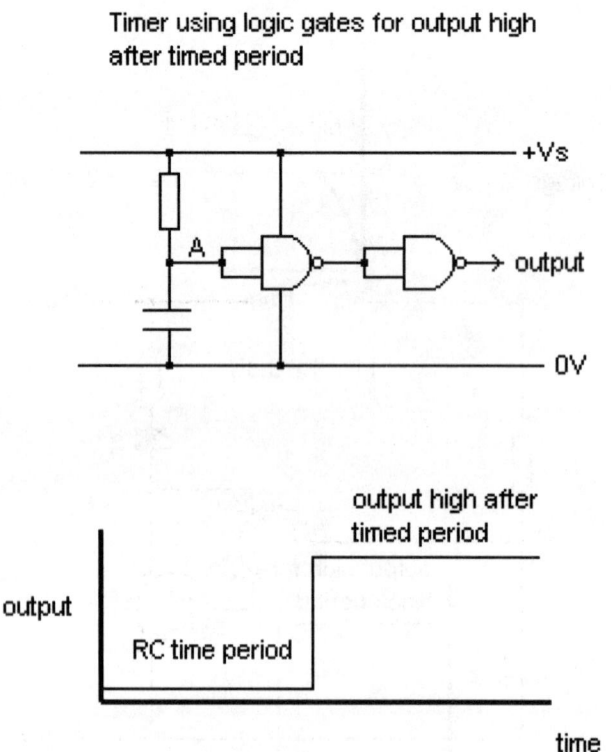

FIGURE 3.4a and b

To overcome the problem mentioned previously regarding large value resistors' inability to allow sufficient base current to turn on a transistor, it is possible to use a Darlington Pair, or a Darlington transistor array, to amplify the current, as shown in Figure 3.5a. The output of this circuit is shown graphically in Figure 3.5b. A slight variation to this circuit is shown in Figure 3.6a, where upon the circuit's being powered up, the timing capacitor seems to be a low resistance, turning on the transistor array until the capacitor is fully charged and no current flows. The output from the Darlington Timer is shown in Figure 3.6b, illustrating how the relay is activated for a set period after the circuit is powered up and while the capacitor is charging up. The problem with this circuit is that if the capacitor is at all "leaky," sufficient current through it would keep the transistor array switched on permanently.

Timer using Darlington Pair

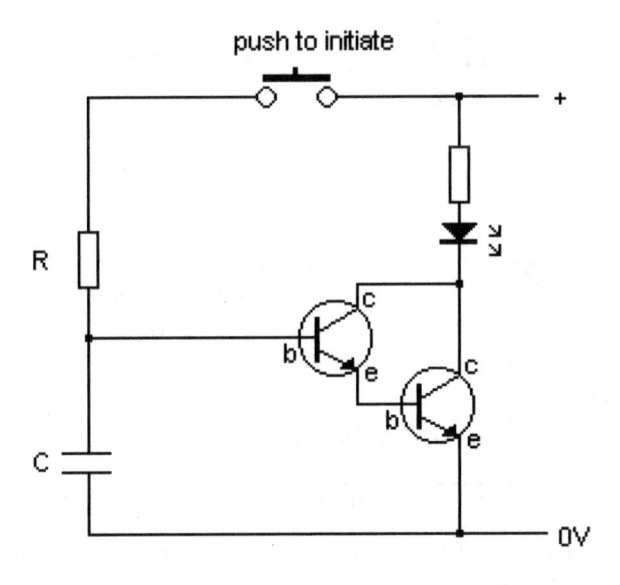

FIGURE 3.5a

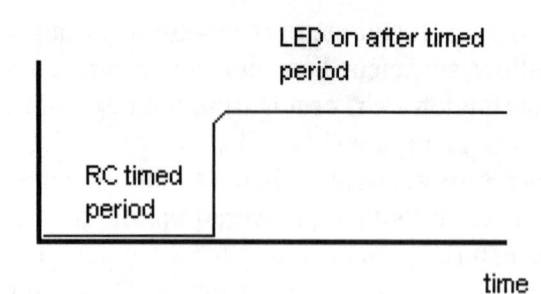

FIGURE 3.5b

Timer with Darlington Pair and relay output

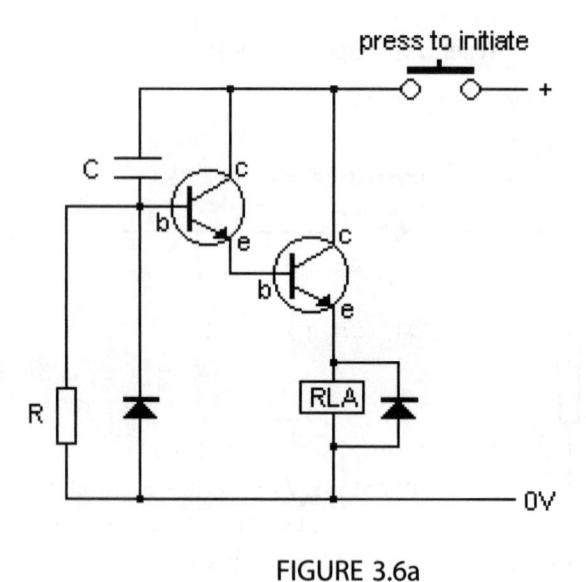

FIGURE 3.6a

A field-effect transistor (FET) may be used with an RC timing circuit, as shown in Figure 3.7. The various biasing resistors may be selected so that the FET is not switched on until, say, half the supply voltage is reached. Besides the fact that the FET has a high impedance, it is also a popular choice for RC timing circuits.

relay activated for RC period

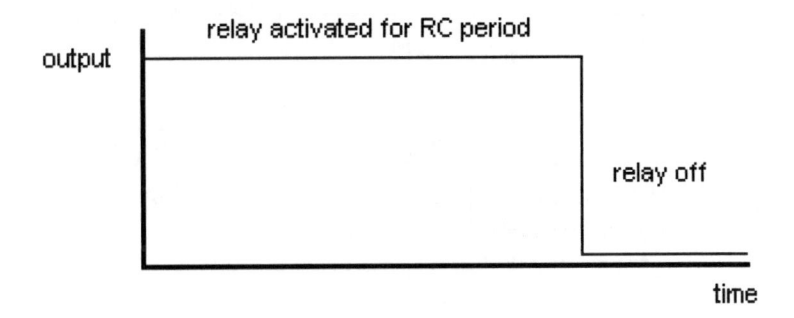

relay off

FIGURE 3.6b

Timing circuit using a FET

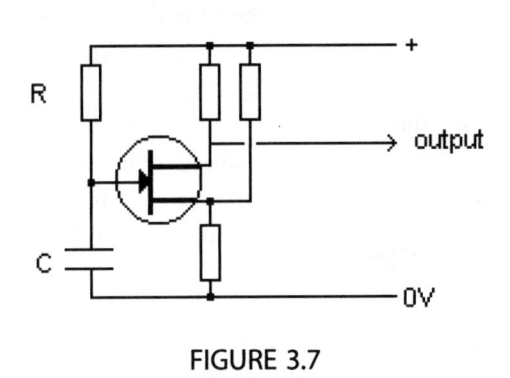

FIGURE 3.7

Figure 3.8 shows an op-amp being used in a timer circuit. The resistor network that is connected to the inverting input can be calculated so that the noninverting input voltage has to reach a relatively high voltage before the output of the op-amp switches to high, thereby providing relatively long time periods when compared to the bipolar transistor timing circuit.

The 555 Timer Integrated Circuit

There are a handful of integrated circuits (ICs) available on the market whose specific purpose is to act as a timer, and the most popular and common must be

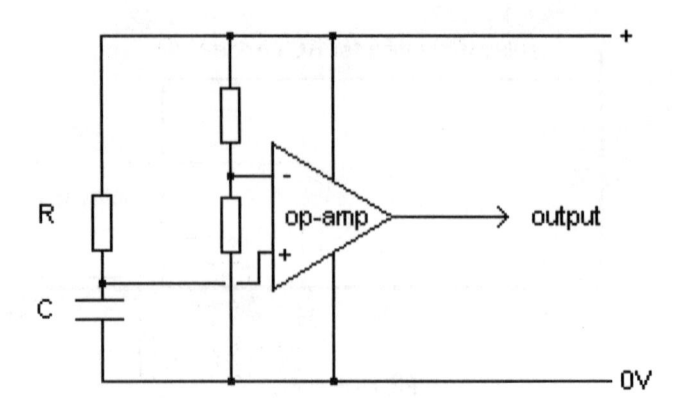

FIGURE 3.8 Timer circuit using an op-amp

the 555 timer chip. Because this IC probably is the most documented of all time, we will not concern ourselves with the internal workings of this device but will, in the next few chapters, concentrate on some practical applications of the 555 timer. Figure 3.9a shows the 555 being used as an RC-controlled timing device. The output, pin 3, will be low until a negative-going trigger pulse is received at pin 2, the trigger input. The output pin will now go high for a period of 1.1xRxC, as shown in Figure 3.9b. The output remains high until the voltage on pin 6 and pin 7 reaches approximately two-thirds of the supply voltage to the chip.

For working out the resulting time for various combinations of R and C, where the formula T=1.1xRxC is true, such as in the case of a 555 monostable circuit, please refer to Table 3.1.

TABLE 3.1 Equal Mark/Space Ratio Astable Multivibrator

(formula = period = 1.4RC or frequency = $\frac{1}{1.4RC}$)

	0.001uF	0.01uF	0.1uF	1uF	10uF	100uF	1000uF
1k		70kHz	7kHz	700Hz	70Hz	7Hz	1.4S
10k	70kHz	7kHz	700kHz	70Hz	7Hz	1.4S	14S
100k	7kHz	700Hz	70Hz	7Hz	1.4S	14S	140S
1M	700Hz	70Hz	7Hz	1.4S	14S	140S	1400S
	frequency			period			

Timing circuit using the 555 ic

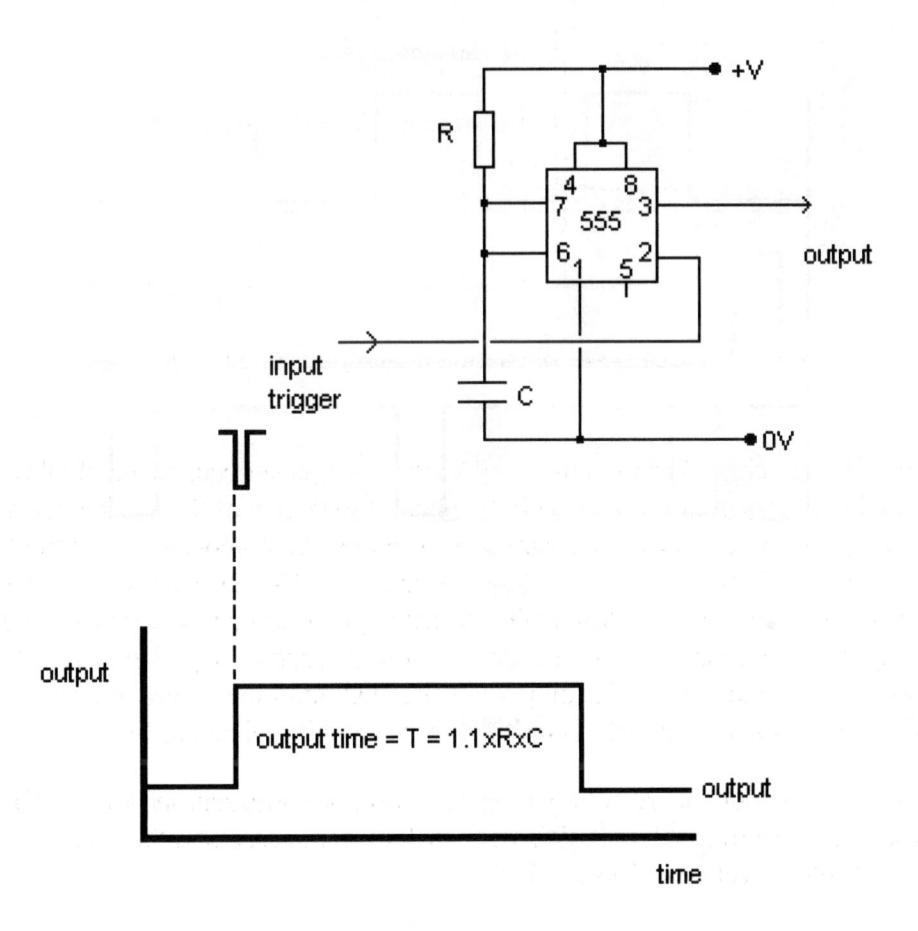

output time = T = 1.1×R×C

FIGURE 3.9a and b

The 555 timer may also be configured as an astable multivibrator. In this configuration, the output toggles on and off, with a waveform that may have an equal mark-to-space ratio; the output is on or off for an equal length of time, on for a longer time than the off period, or vice versa. Typical waveforms are shown in Figure 3.10, with all three examples mentioned. Figure 3.11 shows the 555 timer configured as an astable. The required formulas for calculating both on time and off time are as follows:

Typical waveforms generated by an astable multivibrator circuit

equal mark/space ratio

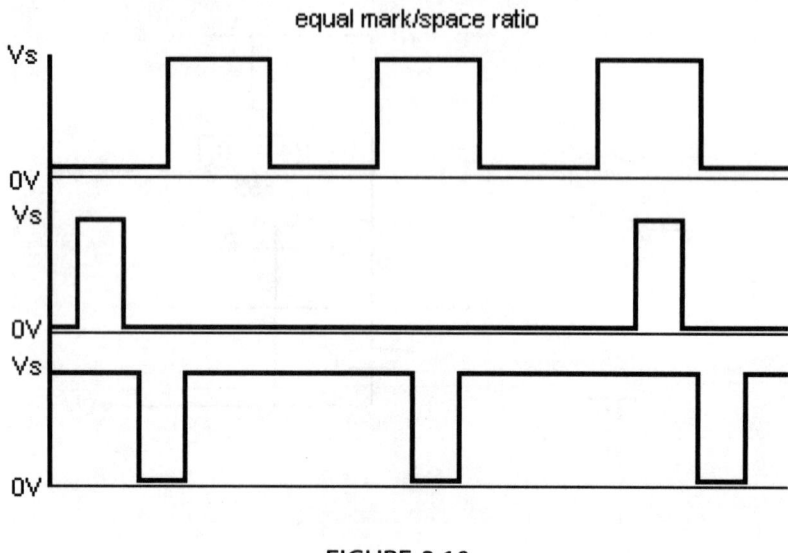

FIGURE 3.10

Astable multivibrator using the 555 timer ic

time on = 0.7 × (Ra+Rb) × C

time off = 0.7 × Rb × C

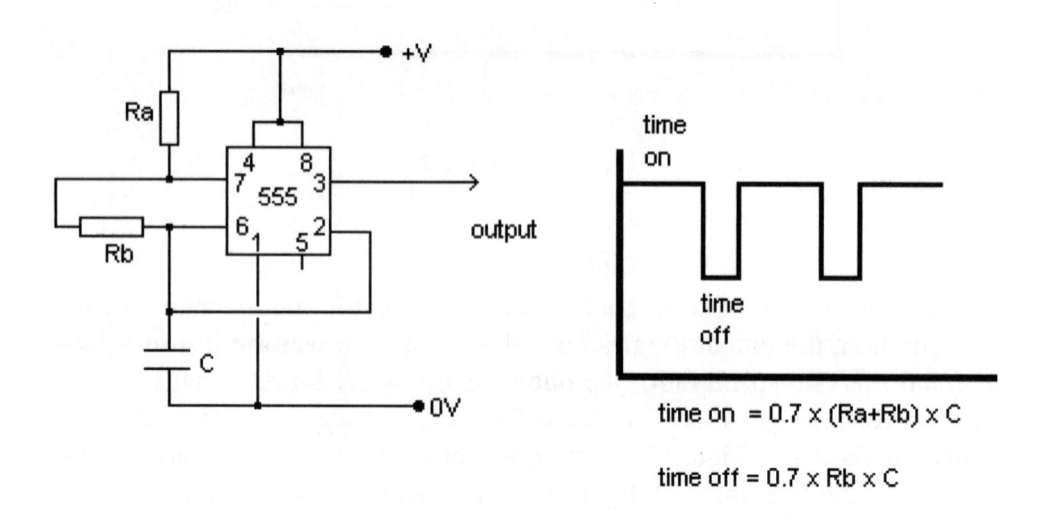

FIGURE 3.11

TIMING CIRCUITS 49

$$\text{Time on} = 0.7 \times (Ra+Rb) \times C$$

$$\text{Time off} = 0.7 \times Rb \times C$$

where R is in ohms and C is in farads.

An equal mark-to-space ratio oscillator using the 555 is shown in Figure 3.12, where the component values of R and C control the frequency of the oscillator:

$$\text{Period } T = 1.4 \infty R \infty C$$

where R is in ohms and C is in farads.

It may be necessary to drive a large load from the 555 timer IC. Unfortunately, the 555 is capable of sourcing or sinking a maximum current of only 200 mA,

Equal on and off times with the 555 ic

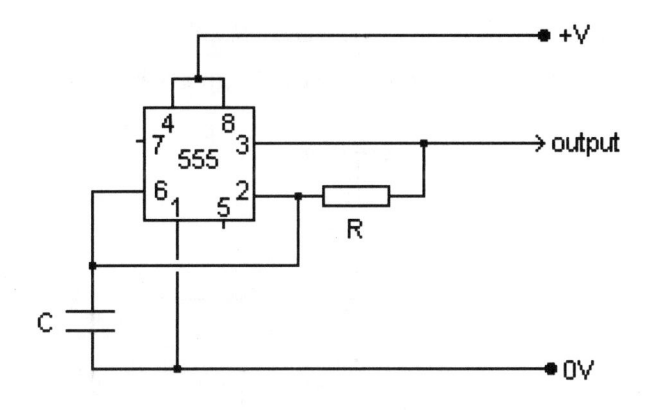

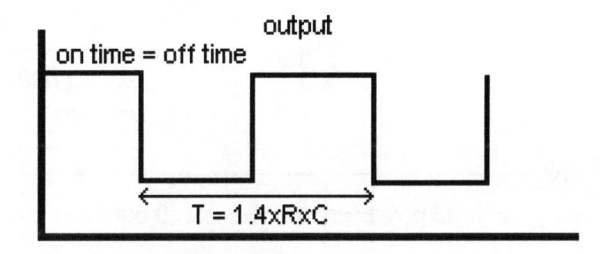

FIGURE 3.12

which is sufficient to drive only relatively light loads such as low-level speakers and the like. This low-level drive may be increased by connecting the output pin of the 555, pin 3, to a transistor that in turn is connected to a relay, which has sufficient contact ratings so as to switch the required load. A circuit using an npn transistor driver is shown in Figure 3.13a, whereas a circuit using a pnp transistor driver is shown in Figure 3.13b.

Method to enable the 555 ic to drive a large load

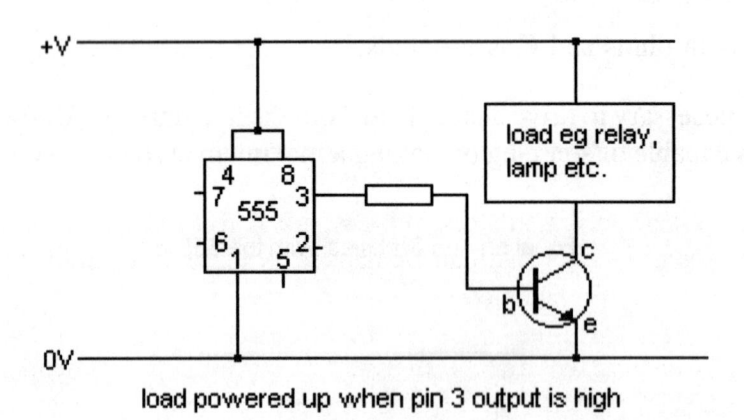

load powered up when pin 3 output is high

Figure 33a

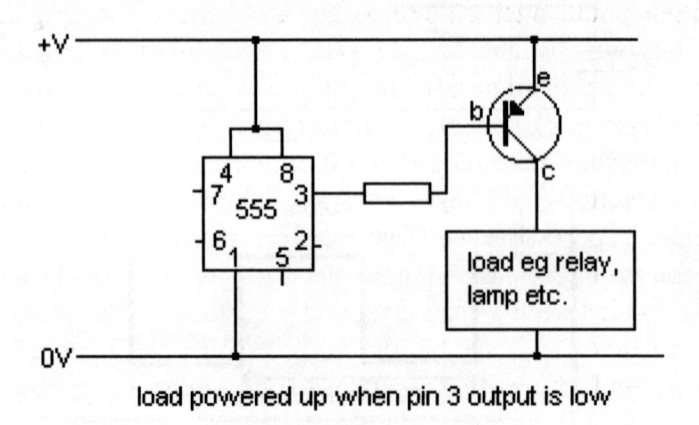

load powered up when pin 3 output is low

FIGURE 3.13a and b

CHAPTER 4

Circuit-Protecting Devices

This is not a note on circuits that protect, but is intended to let the designer be aware of the damage that can be caused to his or her circuit by various problems that may be encountered.

The use of transient voltage suppressors is strongly recommended if economically feasible. These suppressors protect the input and output circuitry from being damaged by short-duration, high-voltage pulses that may occur on these lines from time to time, caused by lightning, induction from high-powered mobile transmitters, and so on. Typical points on a security system that may require safeguarding are shown in Figure 4.1. These points include all exit/output points such as bell, siren, and battery-charging connections, plus any input points such as zone inputs, power input, and the like. Circuits that connect to a telephone also may be in need of protection. A standard master telephone socket contains protection circuitry, but any other line such as an extension may require additional circuitry for preventing spikes causing interference or damage. The first enlargement in Figure 4.1 shows how transient suppressors are connected as close as possible between an input and the ground plane, whereas the last enlargement shows the use of a spark gap, which in this case is etched onto the printed circuit itself. The reason for a spark gap is so that the voltage spike can jump across the gap and go to ground, preventing damage to the circuit components. For this reason, the spark gap must be as physically small as possible and preferably be made from etched points or spikes, as illustrated.

An ideal solution to provide an interface that is capable of providing immunity to transient surges is the popular opto-coupler, which is capable of providing isolation of typically 4kV and is useful for providing a barrier between the control panel and telephone line. Figure 4.2 shows a typical opto-coupler device. The input signal feeds the built-in LED, which then optically transmits a switching signal or serial data to the optical receiver (at the time of this writing, with a switching speed of up to five microseconds).

Many circuits use bells and relays, both of which are comprised of a coil of wire. A coil of wire is inductive in nature, which means that when the coil is switched on and off, when the magnetic field collapses, a very large "back emf" (electromotive force) is produced. A coil with a supply voltage of just a few volts may produce a back emf of hundreds of volts, depending on the switching speed, size of coil, current, and so on, so this large voltage may cause damage to, say, a transistor that is driving the inductance. To minimize the damage that may be caused, diodes are used to limit the back emf voltage to around 0.7V by connecting the diode in parallel with the inductive component, as well as occasionally across the driver device, as shown in Figure 4.3. Each of these four circuits contains an inductance and therefore requires protection diodes as discussed.

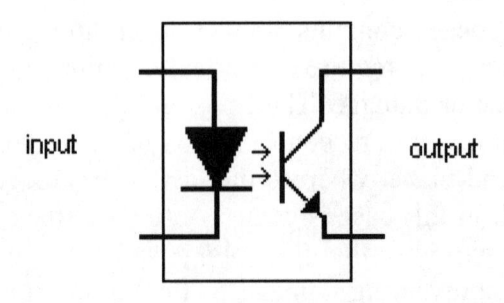

FIGURE 4.2 Typical optocoupler

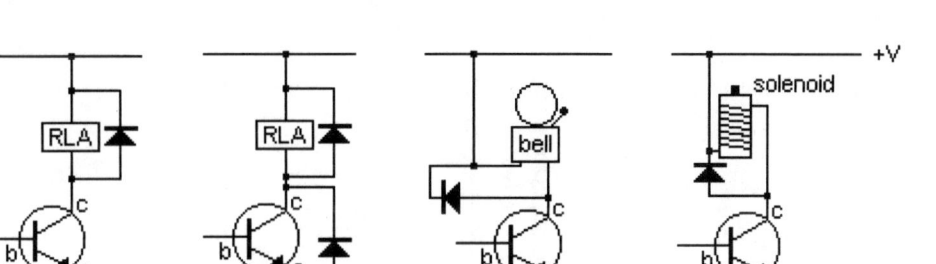

Examples of using diodes to protect against back emf

FIGURE 4.3

If a circuit is DC-powered and occasionally will be disconnected from a supply, for instance if the backup battery needs servicing, it is possible for the supply to be connected with reversed polarity by mistake. To prevent damage being caused to a valuable circuit, you can use a diode or two for protection. Figure 4.4a shows a single diode connected in-line with the supply, so as to prevent any current flowing if the supply is reversed. In the case of Figure 4.4b, two diodes are used. D1 performs the same function as in the previous example, but D2 performs the task of disabling the circuit if the latter is powered up from a reversed supply. If the supply line is backward, D2 will conduct, thus passing excessive current through the fuse, F1; as long as the power supply can provide sufficient current, both fuse and diode will be damaged, the latter being especially true if the fuse has a delay action!

This "suicide diode" was often used in CB (citizens band) radios and provided a steady "bread-and-butter" income for many technicians in bygone days, because all that was required to fix the rig was a fresh fuse and a diode. The suicide diode may be used in several designs today; protecting the circuitry from damage by the consumer's simply replacing the blown fuse with a much heavier fuse and connecting the supply backward again will, in theory, simply blow the fuse again, just so long as the suicide diode actually has been blown into short-circuit mode. If the diode has been blown but is open-circuit, as occasionally occurs, the theory of this particular

54 APPLIED SECURITY DEVICES AND CIRCUITS

Circuit protection diodes

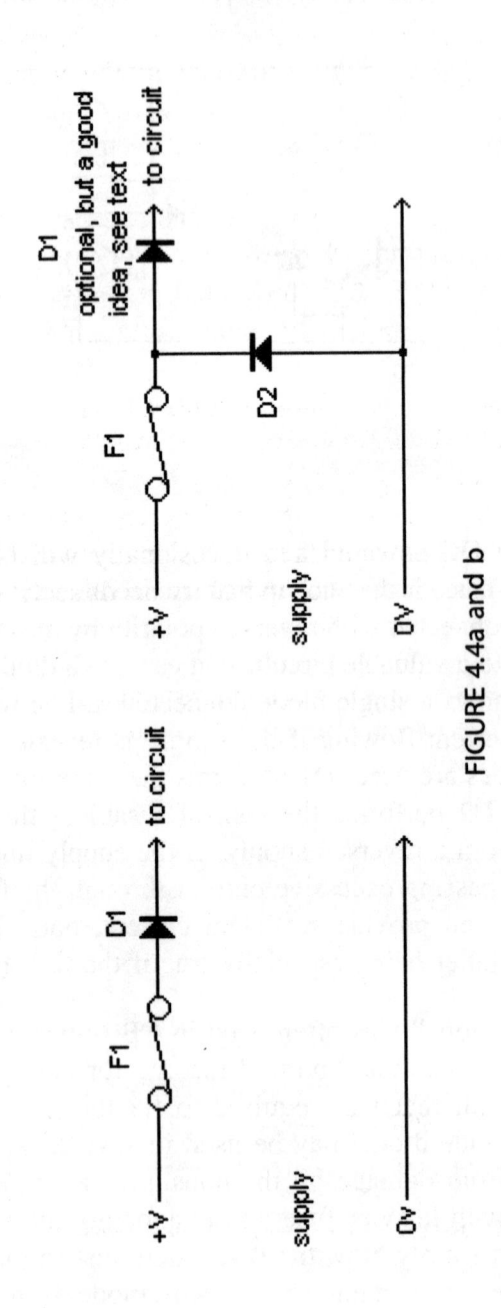

FIGURE 4.4a and b

circuit would not work were it not for the presence of D1, which ensures that the circuit will not power up due to the DC blocking action by this diode (illustrated in Figure 4.4b). Unfortunately, the circuit being supplied by this supply may have to be redesigned somewhat due to the small voltage drop across D1, which would be around 0.7V.

Because of the nature of security systems, measurement circuits and remote-control circuits that are all hard-wired, very long runs of cabling may be required to connect the control panel or processor to the sensor. If economically possible or if called for, connecting cable may be of the screened variety; otherwise, an unscreened cable consisting of twisted pairs may be used. The reason for this cautionary note is that electrical interference may well be picked up by the cable en route, from adjacent mains cables, electric motors, radio transmitters, cellular or cordless telephones, televisions, and the like. This interference then will be transported to the input circuitry, which may be a signal amplifier, triggering a false alarm. This problem may be minimized by the judicious use of small-value capacitors, perhaps around 1nF, that effectively will short out the interfering signal to ground.

Typical points in a circuit where these decoupling capacitors may help are seen in Figure 4.5. Capacitor C3 probably will be already included within the

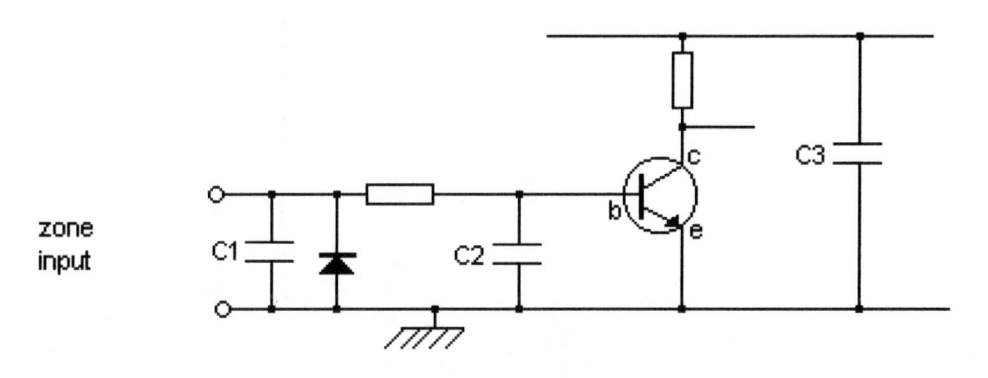

FIGURE 4.5

existing circuit across the output of the voltage regulator, though it may be prudent to add further decoupling capacitors in close physical proximity to the processor, amplifier, and so on. Components C1 and C2 may have to be chosen carefully, if not excluded in some designs, if their presence has an adverse effect on wave-shaping and the like.

CHAPTER 5

Automobile Security

When considering installing some kind of vehicle security, or improvements to vehicle security, think of the automobile crime reports you read in your local newspapers. Large numbers of thefts from and of automobiles, as well as damage to vehicles by vandals or attempted break-ins, apparently are recorded every hour—sometimes every minute—of every day.

We have no surefire methods of protecting our vehicles, but it is possible to provide a high degree of deterrence. Visible alarm indicators and security stickers on windows may encourage the average thief to move on to a vehicle without these deterrents, thus leaving your vehicle intact. However, if you are foolish enough to leave an expensive jacket in view, instead of locking it in the trunk (before the start of the journey, considering that thieves hang around in parking areas to see what you are hiding), or even leaving an empty pack of cigarettes on the dash, then you are leaving yourself vulnerable to getting a broken window.

Factory-fitted steering locks on certain types of vehicles may be flimsy and easily forced. The cheaper types of steering-wheel locks you can purchase on the market give the thief more leverage than having to overcome the factory-fitted lock, after which the thief removes or defeats the cheaper device in one way or another. Purchase an insurance-approved device, and/or fit an engine immobilizer—or three—to your vehicle.

The favorite target of thieves is the vehicle radio, although you would think that so many are stolen and sold for such paltry sums that legitimate radio vendors would be out of business by now. Unless the radio is installed with a quick-release system, enabling the radio to be quickly removed and concealed under a seat, for example, our only course of action is the visible deterrents mentioned previously, plus having coded radio stickers on display.

AUTOMOBILE CIRCUITS

This section presents ideas and designs for automotive security, automobile protection, and monitoring automobile systems.

Automobile/Backup Battery Monitor Circuit

This circuit, shown in Figure 5.1, is extremely useful for monitoring the 12-volt batteries commonly found in automobiles, vans, and boats, or as backup batteries in security alarm systems. This circuit is particularly useful if an automobile does not have any other indicator to show that the alternator/battery-charging system has stopped working, or if the belt has broken, since this is not obvious until a breakdown occurs, which could be a few days after the event. The circuit also may be used in conjunction with a commercial or homemade battery-charger unit.

This circuit has an attractive display indicator that looks (almost) like traffic-control lights, using a red, yellow, and green array of LEDs. A table showing the LED output for various voltage inputs is shown below.

$$RED < \ = 11.6V$$
$$RED+YELLOW = 11.6V \text{ to } 12.0V$$
$$YELLOW = 12.0V \text{ to } 12.7V$$
$$GREEN+YELLOW = 12.8V \text{ to } 13.2V$$
$$GREEN> \ = 13.5V$$

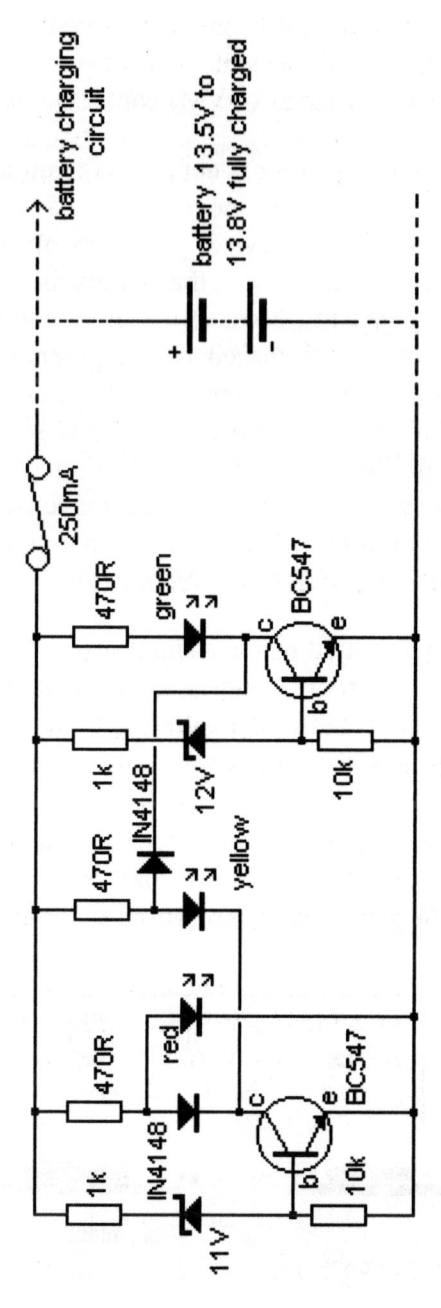

FIGURE 5.1

Note that these values are approximate only, due to the tolerance of the components used. To calibrate the unit, it is preferable that the circuit be connected to a laboratory power supply with a variable output over the desired range and with a high-impedance digital volt meter (DVM) connected across the terminals.

Do not be too concerned if the unit is connected to an automobile and shows only a "perfect green" while on the move. If you have your electrical system checked out after installing the device, you probably will find that a green indicator is possible to obtain only if the battery is fully charged and connected to the running alternator. Most of the time, you will observe a yellow indication while the engine is turned off, a green indication and yellow indication combination while the engine is idling, and a red indication and yellow indication combination while the engine is somewhere between idle and stall or idle and all electrical items switched on. A red indication should be noted only if the engine has stalled, with all electrical items switched on. If you are now totally confused, I recommend that you build this circuit and try it out for yourself.

The whole circuit may be fitted inside a small plastic "potting box," and then, after a small hole is drilled in the vehicle dash at a suitable point, the two wires to the supply can be connected to the battery circuit. The box can be secured either with self-tapping screws or stuck down with double-sided adhesive tape, as shown in Figure 5.2.

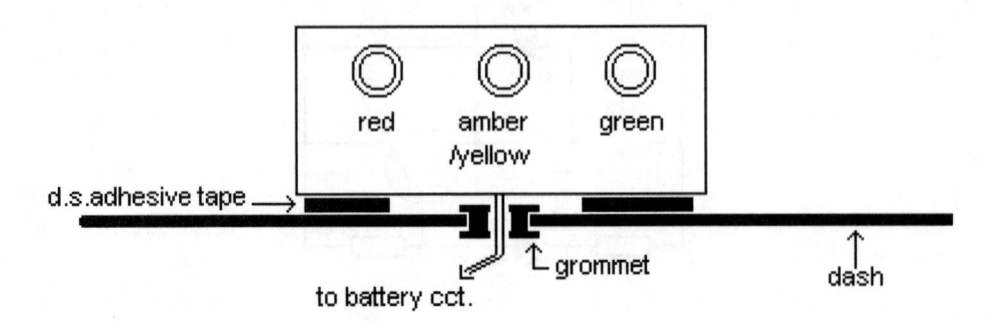

FIGURE 5.2

It is a good idea to fit a small quick-blow fuse in-line with the supply from the battery to limit the possibility of serious damage. Do not do what I did when using the device for the first time, that is, sit at traffic lights lightly revving the engine and watching the LEDs change color, at the same time not noticing that the real traffic lights had gone through their cycle twice! It may be a good idea to point the LED display away from the driver's view if he or she is easily distracted.

Remote Key Fobs

Remote key fobs are supplied with many automobile alarm systems. If a bunch of keys are lost or stolen, there is a good chance that a) the bunch contains the vehicle key, b) the bunch contains the alarm fob, c) you have a spare key at home that can be sent over to you, and d) you never thought to get a spare alarm fob. Unfortunately, the price of the option of a second "spare" fob sometimes can be almost as much as the complete system. If a redundant key fob is handy—perhaps your old alarm is the same type as the newly purchased one—it may be possible to "clone" the new one, thus generating a spare fob at no extra expense.

Many commercial automobile alarm systems in the lower price bracket use a chain of coded transmission from the key fob, which uses the same code on every press, whereas more complex systems transmit a different code every time the alarm is used and so are excluded from this option. If the alarm system is of the first type, the new fob can be opened up and the code seen from the way the encoding IC is wired (see Figure 5.3). These encoding chips have a tri-state input, so they can produce a much greater range of codes; only two states would give relatively few choices. The IC may have an input connection that is connected either to the positive rail or the negative rail, or is left disconnected, that is, floating.

Copying the code to an old key fob may entail no more than unsoldering and soldering a few solder bridges, or at worst, cutting tracks with a scalpel and bridging cut tracks. The two ways to bridge a cut track are either to carefully scrape off the etch resist paint and, after tinning both clean track ends, use a blob of solder or solder a thin length of tinned copper wire across the gap, or

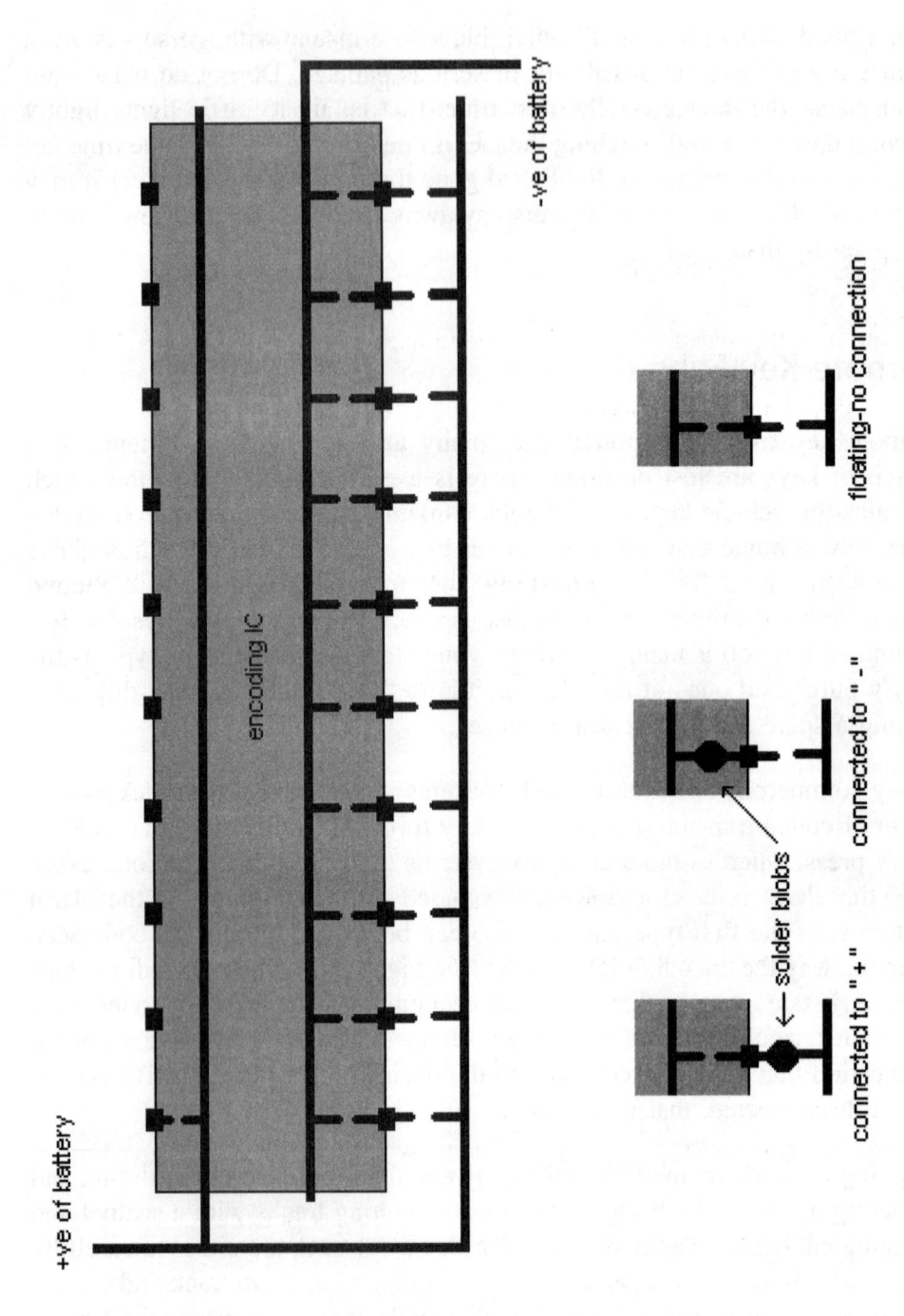

FIGURE 5.3 Tri-State Remote KEYFOB Settings

alternatively—and probably the best method—to obtain a length of thin pvc-insulated wire and route this from each respective solder pad. The pvc-insulated wire then can be tidied up, painted, and glued to the fob pcb (printed circuit board) with a proprietary pcb repair paint.

Lights-On Reminder

Although some automobiles have an audible alarm to warn drivers that they have left their lights on, in some automobiles, this device either is not present or has an insufficiently loud warning. The circuit shown in Figure 5.4 has a minimal component count and may be assembled in less than one hour.

Connections for the alarm are as follows (please refer to Figure 5.4a):

- ▲ Point A is connected to a point that is +12V when the car lights are switched on.

- ▲ Point B is connected to a point that is +12V when the ignition is switched on.

- ▲ Point C is connected to a point that is permanently grounded.

Automobile "lights on" reminder

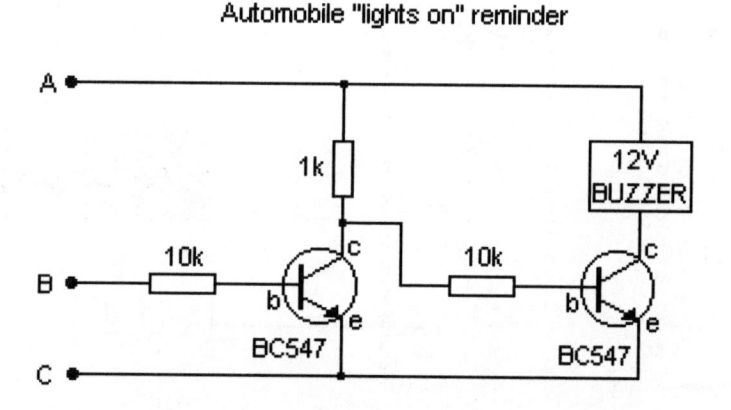

FIGURE 5.4

64 APPLIED SECURITY DEVICES AND CIRCUITS

Lights on reminder alarm wiring details

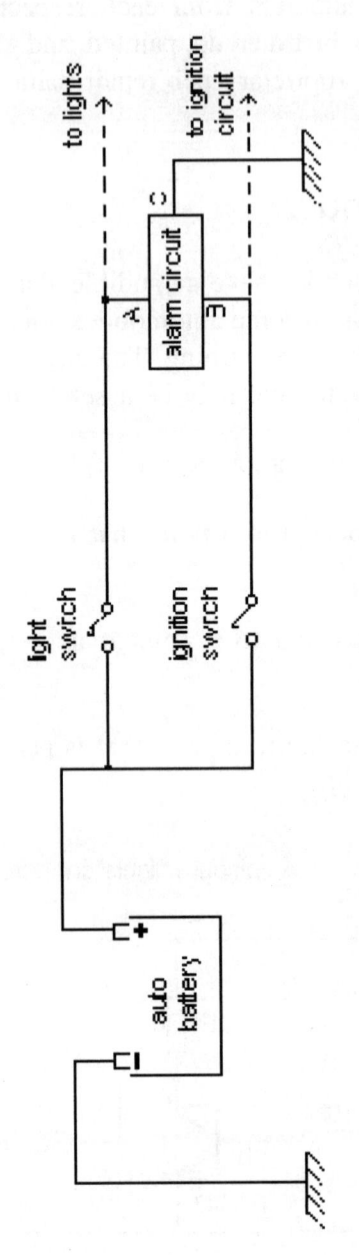

FIGURE 5.4a

The best point to find these switched feeds may be at the fuse box, which, if inside the vehicle, saves having to route wires through the bulkhead. Remember that if this must be done, then well-sheathed wire and suitable insulative grommets should be used in order to prevent chafing and cutting of the wiring. Also be careful not to foul up anything important with your wiring, such as brake pedals!

If the lights are switched on and the ignition connection is live, the buzzer will not sound, because the first transistor is being held on by the ignition voltage; therefore, the base of the second transistor is effectively at ground potential. If the ignition voltage now is removed, the second transistor turns on as the collector voltage of the first transistor goes high, so operating the buzzer. A table showing this operation follows.

Ignition Input (point B)	Lights Input (point A)	Buzzer
0	0	0
1	0	0
1	1	0
0	1	1

Automobile Headlight Sensor

It is very inconvenient to have a headlight on a vehicle burn out, especially since the first time you may be aware of the fact is when stopped by a policeman who hands you a ticket on the spot—or inspects the vehicle and finds other things wrong! If the headlight is not the sealed type, it may be possible to drill a hole somewhere in the headlight enclosure and then mount the LDR as flush as possible, so as not to unduly interfere with the lamp pattern. See in Figure 5.5 that, when the lamp is illuminated, or during daylight, the resistance of the LDR falls to a minimum value, effectively shorting out the LED. If the lamp is burnt out, then the LDR is of maximum resistance, which means that the potential difference across the LDR is greater, thus illuminating the LED as current passes through the 1k resistor.

Automobile lamp remote sensor

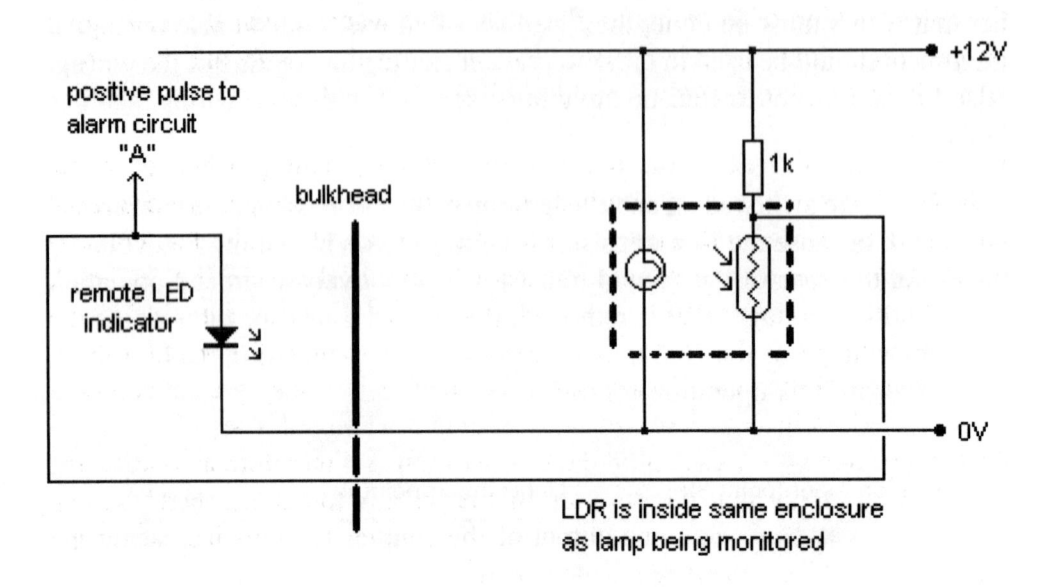

FIGURE 5.5

This concept is in no way limited just to automobile headlights but can be used whenever necessary to remotely monitor other important circuits. If point "A" is taken to an alarm circuit, when the sensor is activated, this point becomes more positive so could act as a trigger input.

Automobile Indicator Alarm

Every motorist is aware of the problem on the roads today of persistent "turn signal–itus," where the driver in front indicates for miles that he or she intends to turn but never does. This means that other drivers at intersections pull out in front of the car in front, you are afraid to pass in case the offender does actually turn, and so on. The reason for persistent turn signal use usually is twofold: 1) Indicators of the self-cancelling variety require a sharp opposite locking of the steering wheel to cancel, but if the maneuver is a gentle one, the self-cancel does not always work, and 2) the indicator alarm is reliant on the quiet clicking of the relay to give an audible warning that the turn signal is operating. True, there is a visual indication on the dashboard that a turn signal is in use, but nobody's perfect!

It is possible to enhance the noise of the relay by adding a buzzer to the indicator, but when waiting at a busy intersection, with temper or adrenaline increasing by the second, a buzzing with every click of the indicator will only make matters worse. The solution to this problem is shown in Figure 5.6. A decade-counter IC, the 4017, is used to divide the clocking signals derived from the existing indicator relay by a factor of 10. Circuit isolation is provided by a diode, and then the clock signal is clipped to 12 volts by a zener diode. Once 10 pulses have been received, the output at pin 11 goes high for the duration of one pulse, switching the transistor on and therefore activating the buzzer. If you find that the duration of the buzz is too short, then you can always connect more than one output of the counter IC, thus increasing the effective length or repetition of the sound.

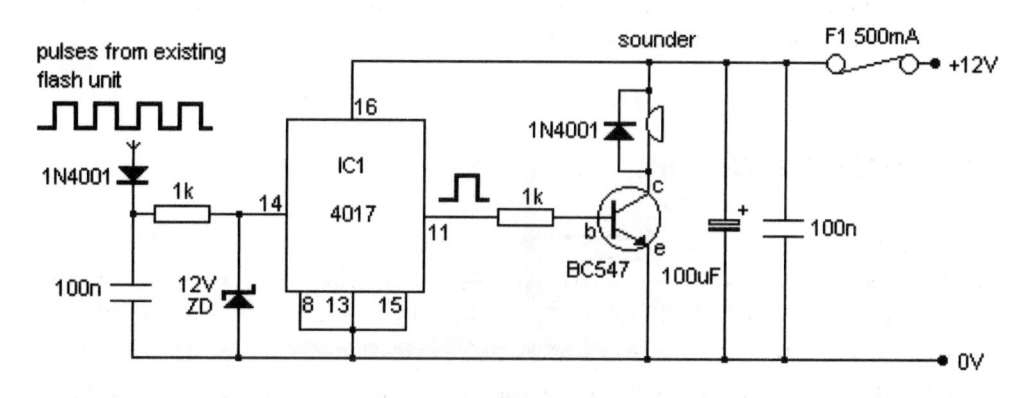

FIGURE 5.6

Vehicle Knock Alarm

This device produces a one-off, short-duration, audible alarm if the vehicle is knocked by vandals or by another vehicle. The circuit shown in Figure 5.7 is based upon the vibration alarm described in Chapter 8, "Miscellaneous Designs." The time for the duration that the sounder is on is adjusted by the variable resistor. When the trembler switch is activated, the electrolytic capacitor becomes charged up instantaneously and then starts discharging through the Darlington Pair, with the speed of discharge controlled and limited by the variable resistor. While the capacitor is discharging, there is sufficient current to keep the transistors on; therefore, the relay remains activated for the preset time.

If a commercial trembler switch is unobtainable, you can fabricate one using the design shown in Figure 5.7a. You can build a sensitivity control into the design by placing an adjustable screw through the upright board, so adjusting the distance between the suspended weight and screw contact. To avoid having to modify the existing horn circuit of the automobile, of which there are many designs, it's prudent to use a separate siren, which probably would be more noticeable anyway. The two separate transistors may be replaced by an integrated Darlington device. If the time period is too short for any particular

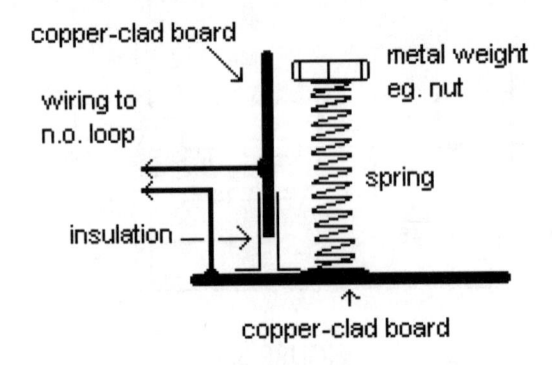

FIGURE 5.7a

application, then increase the value of the leakage-control resistors—a cheaper method than increasing the size of the timing capacitor. If the electrolytic capacitor is increased to a very large value—in the region of thousands of microfarads—the trembler sensor switch contacts may spark on initial contact, causing corrosion, or the contacts may even fuse together, because the initiating charge current is limited only by the internal resistance of the battery.

Cigar Lighter Thief Confuser

This design provides an effective method of fooling a potential automobile attacker into thinking that the vehicle is alarm-protected. The unit, shown in Figure 5.8a, consists of a dual LED flasher circuit housed inside a standard

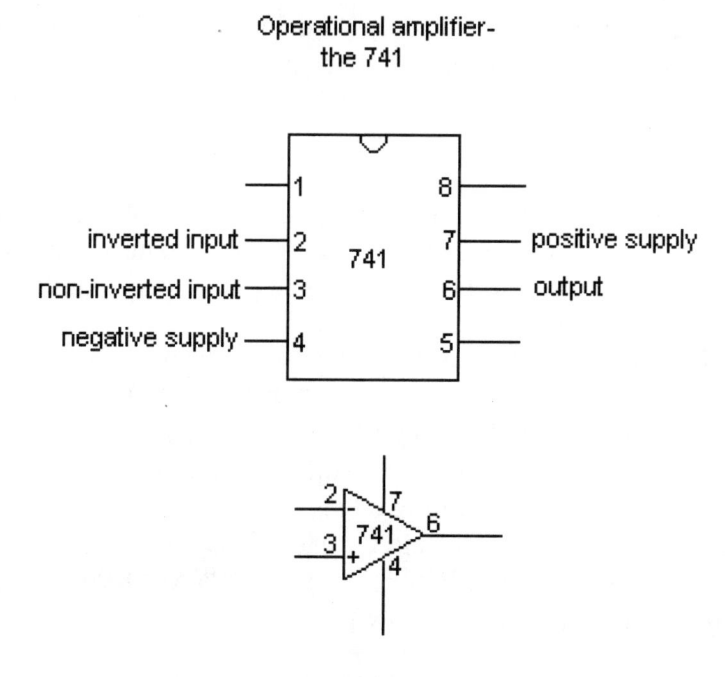

Operational amplifier-
the 741

pin 1
pin 2 — inverted input
pin 3 — non-inverted input
pin 4 — negative supply
pin 7 — positive supply
pin 6 — output
pin 5
pin 8

FIGURE 5.8a

Alternating R/G flasher

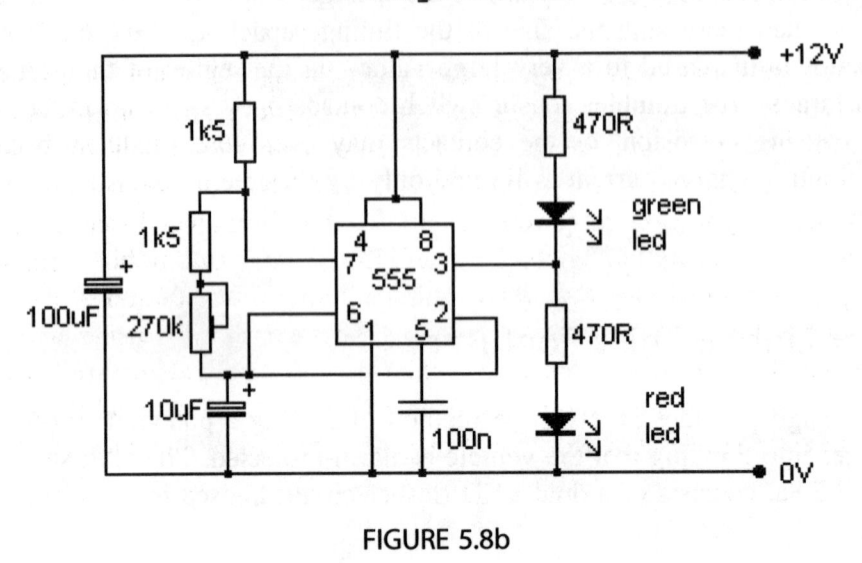

FIGURE 5.8b

cigar lighter plug adapter that is widely available. When the circuit is plugged into the lighter receptacle, it draws power is from the receptacle, therefore avoiding the need for batteries or switches. When connected, the unit produces an eye-catching warning, as the green and red LEDs flash alternately. The circuit for the alternating flasher is shown in Figure 5.8b. Current consumption is low, which means that the unit can be left plugged in for a few days without causing excessive drain to the vehicle battery; however, don't leave the unit on for several days if the battery is in poor condition.

To ensure that the LEDs are clearly visible, use devices of the super-bright or hyper-bright range and, if necessary, ensure that the LED chosen is of wide-angle viewing. If the angle of the cigar lighter receptacle means that the LEDs are pointing at a poor-visibility angle, adjust their mounting angle and glue them into position. If alarm warning labels are affixed to the windows of the vehicle, then the combination of circuit and labels hopefully will make any prospective thief look around for an easier target. (Nevertheless, do not tempt the thief by leaving valuables in view.)

Automobile Headlight Delay

Parking your car at night in an unlit area can be worrying, if not inconvenient. It is useful to have some way of illuminating the parking area for a short while until, for example, your house keys are found, or a villain believes that the automobile is still occupied, perhaps by someone who may prevent a prospective attack on you, so giving you time to reach an illuminated safe area. The circuit shown in Figure 5.9 lets the headlights of your automobile remain on for a predetermined time (about two minutes) after a pushbutton is activated on the dashboard. This time period may be increased by using a higher resistance in place of the 150k, but if the period of time is excessive, battery drain may be a problem. Another problem may exist if the automobile is fitted with a current-sensing alarm, which may be confused by this circuit's being

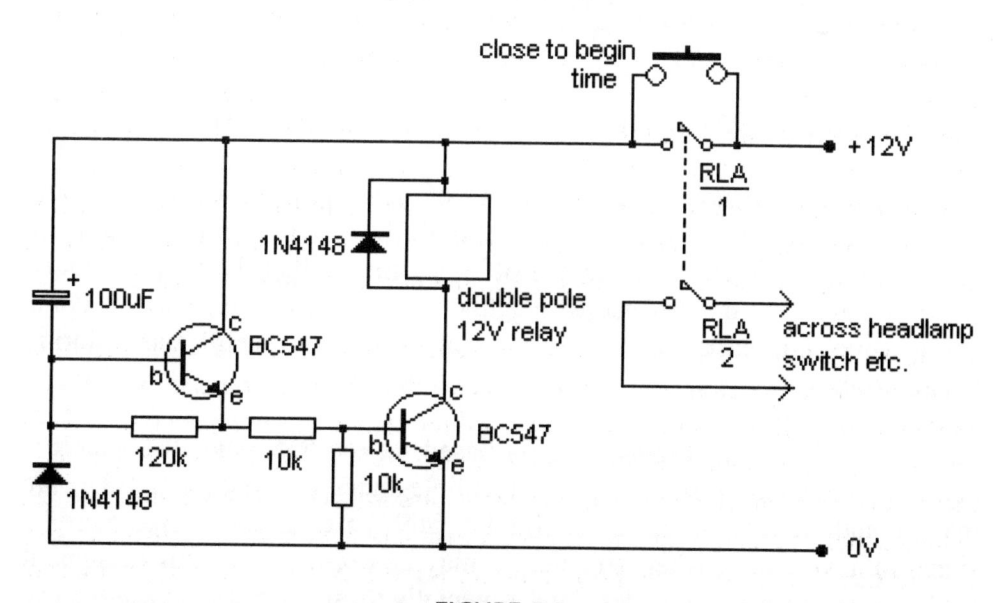

FIGURE 5.9

installed. If all the resistors were stripped from the circuit, this circuit basically is a Darlington Pair configuration. Care must be taken not to exceed the current-handling capability of the headlight-switching relay contact pair on this circuit; a pair of 25W lamps would require

$$W = I \times V \qquad I = W/V$$

$$I = 50/12 \qquad I = 4.17 \text{ amps}$$

therefore, in this case, a relay with a minimum rating of 5A @ 12V would be used.

Another automobile headlight delay circuit is shown in Figure 5.10, this time using the 555 timer IC. When the trigger pin is grounded temporarily using the dash-mounted pushbutton, the output pin goes high for the period set by the RC timing components, in this example, 330k and 100uF. If the time period must be adjustable, then a variable resistor may be connected along with a 100k resistor to replace the 330k. It is important that the relay contacts be of sufficient current rating, as described in the preceding text.

Automobile (Daylight-Only) Reversing Alarm

This design should not be dismissed out of hand as just another reversing alarm, because it is designed to work only during daylight hours, whenever the automobile's lights are switched off. This means that the beeping sound will not cause any disturbance to neighbors if you leave for work or come home while it is dark, but the circuit still provides a useful alarm during daylight hours.

The circuit, shown in Figure 5.11, is based on the 555 timer IC, which is configured in a circuit so as to give an audible signal every second. Take care that the audible warning does not simulate any "safe for pedestrian to cross" signal in any country where this design may be used; for obvious reasons, it could cause confusion to pedestrians, especially those with poor eyesight. The device obtains the 12V and 0V supply lines from the existing reverse warning lamp circuitry and chassis ground, respectively. The circuit will not, however,

Vehicle Reversing Alarm

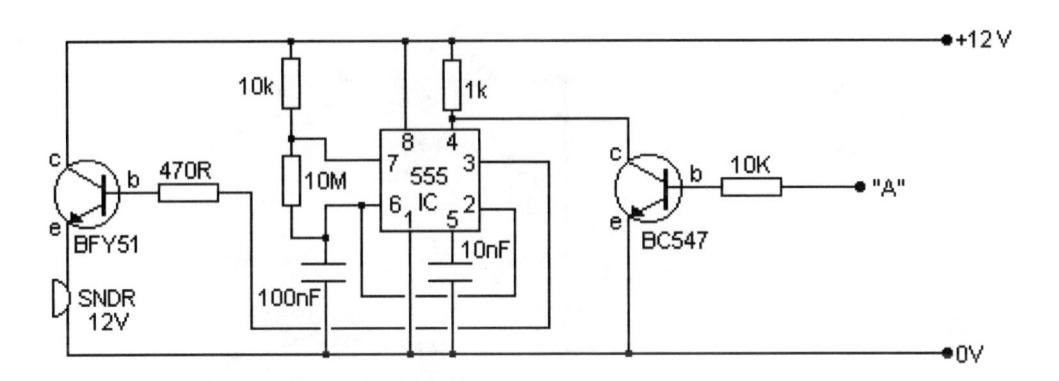

FIGURE 5.11

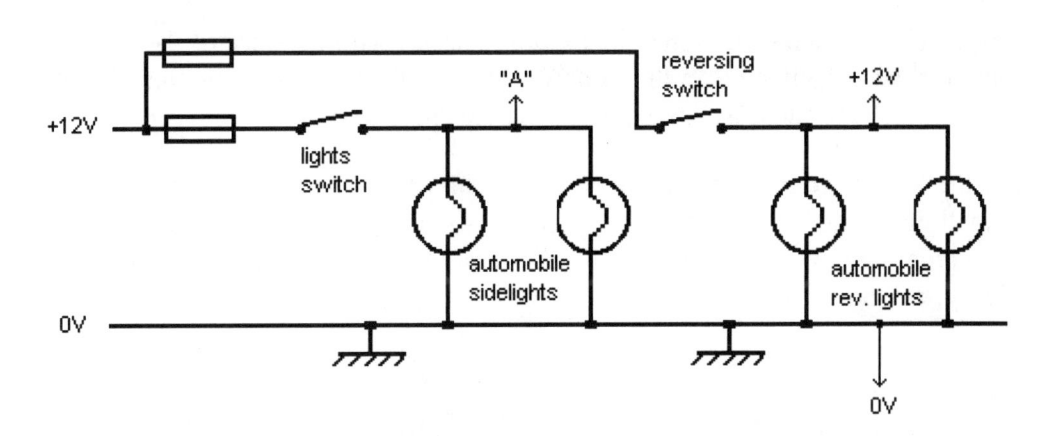

FIGURE 5.11a

produce an output until the transistor—which effectively shorts pin 4 of the timer IC to ground if the transistor is switched on—has the voltage removed from its base. This controlling voltage, fed in at point A, is obtained from the existing automobile sidelight circuitry, as shown in Figure 5.11. When this point is made positive, either by the sidelight switch or associated existing relay contacts (see Figure 5.11a), the transistor is turned on, therefore inhibiting the beeper circuit. If the inhibit system is not required or has to be

Relay output for large load

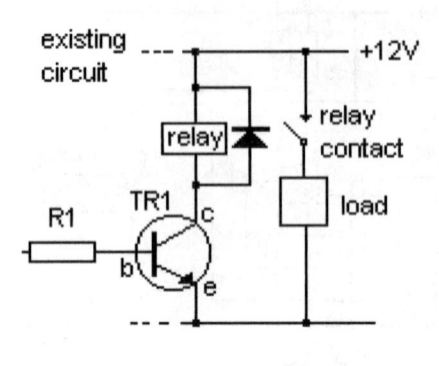

FIGURE 5.11b

overridden, then a switch may be placed in series with the base of the switching transistor, so that when the switch is opened, the transistor remains off and the 555 IC operates, because pin 4 is at full potential.

If the output transistor—in this case, a BFY51—has insufficient current-handling capabilities to drive the chosen sounder, the output transistor can switch a relay whose contacts will have a much larger current-handling capability, as shown in Figure 5.11b.

Automobile Battery Condition Monitor

This circuit, shown in Figure 5.12, uses a 741 op-amp device to monitor the output from an automobile battery. The first design gives a visual output by means of an LED if the battery voltage drops below a preset value as set by the variable resistor. The noninverting input, pin 3 of the op-amp, is set to a definite value by the 5V6 zener diode, at 5.6V! The inverting input, pin 2, is set to a value just above this reference value, so with this more positive than pin 3, the output of the op-amp is off. When the supply voltage falls slightly, pin 3 still is 5.6V due to the stabilizing action of the reference diode, but the voltage divider chain output at pin 2 falls. With the inverting input now more positive

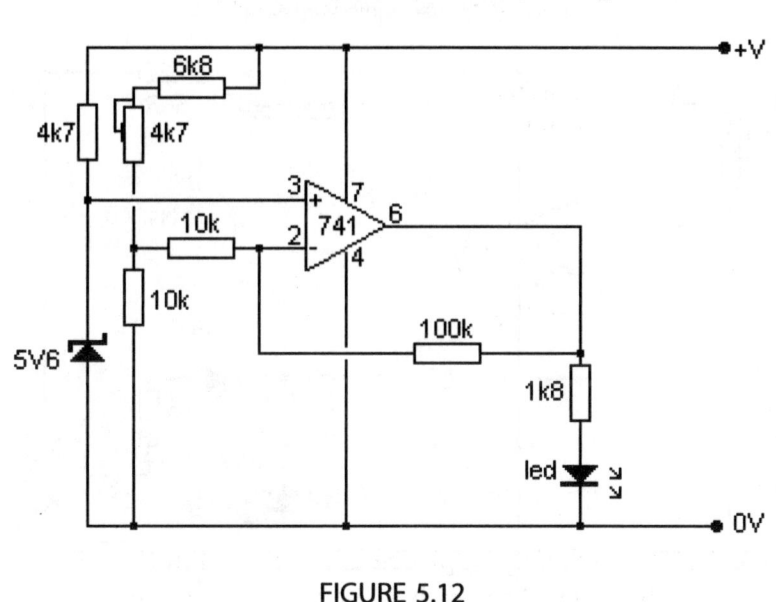

Battery condition monitor using 741

FIGURE 5.12

than the noninverting input, the output, pin 6, now goes high, thus turning on the warning LED. It is easily possible to reverse the inverting and noninverting inputs so that the output is high until the monitored voltage drops down; in this case, the LED warning light could be green for system OK.

The second design, shown in Figure 5.12a, is similar to the first but has an added audible alarm indicator. The transistor used for switching the buzzer on is fed via a 3V3 zener diode, to avoid the problem caused by the LED and its series resistor, which may cause the transistor to turn on annoyingly if glitches appear on the supply line. Glitches may be filtered out by using the 56R resistor and 100uF capacitor, if proving a nuisance.

Automobile "Hot-Wire" Preventer

This design is built around a 555 timer IC configured as an astable, whose output switches the relay on and off in a continuous cycle once activated, and

Battery condition monitor using 741
with additional audio warning

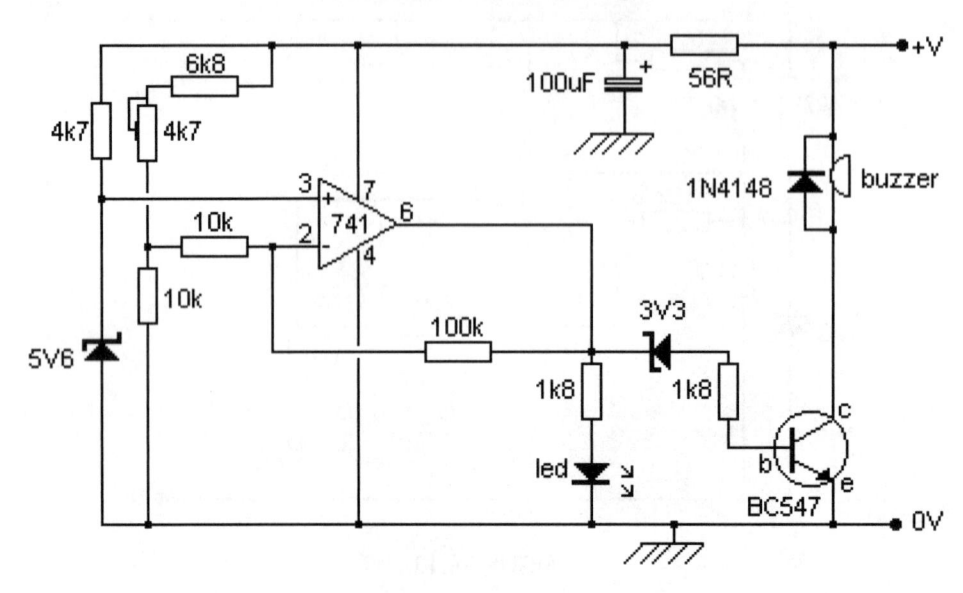

FIGURE 5.12b

is armed by a hidden key switch. The normal method of hot-wiring, shown in Figure 5.13, is to connect the positive terminal of the vehicle battery directly to the input of the ignition coil, thus circumnavigating the ignition switch. This circuit, shown in Figure 5.13a, is based on the 555 timer IC configured as an astable, which runs continuously once armed. When the device is armed, the relay contacts of the circuit will switch on and off, thereby switching the distributor contacts in and out of circuit to ground. This gives the effect of making the automobile run extremely roughly, which should mean that after a few feet, villains will abandon any further attempt at stealing the vehicle, with the vehicle now considered a nonviable asset. The wiring for this "alternative immobilizer" should be very well hidden, with connection to point A being disguised as much as possible and the key switch concealed wherever convenient.

This design should not be dismissed until you consider the amount of disinterest caused nowadays by a vehicle's alarm going off in the middle of the

AUTOMOBILE SECURITY 77

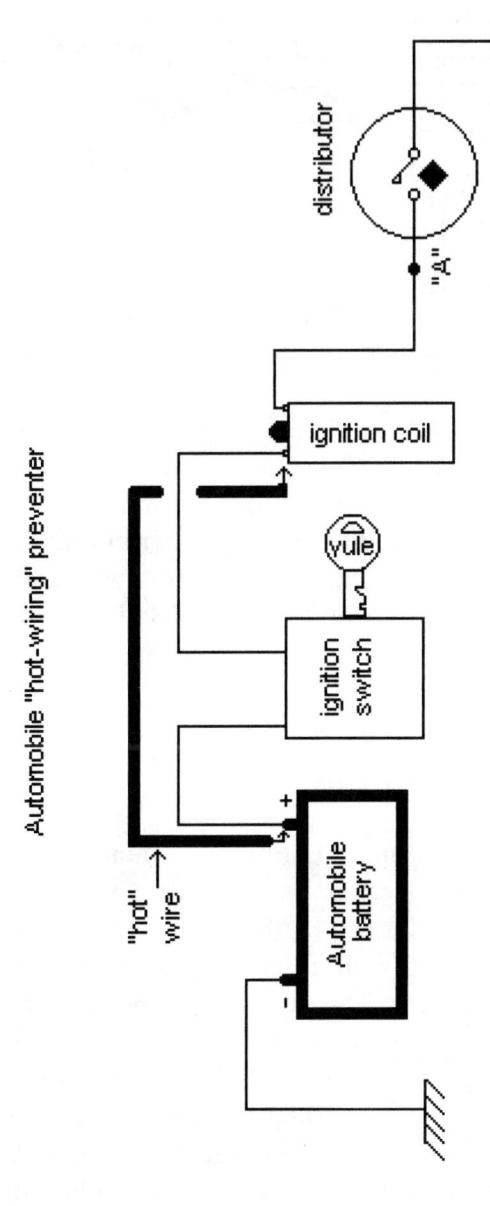

FIGURE 5.13

Automobile "hot-wiring" preventer

FIGURE 5.13a

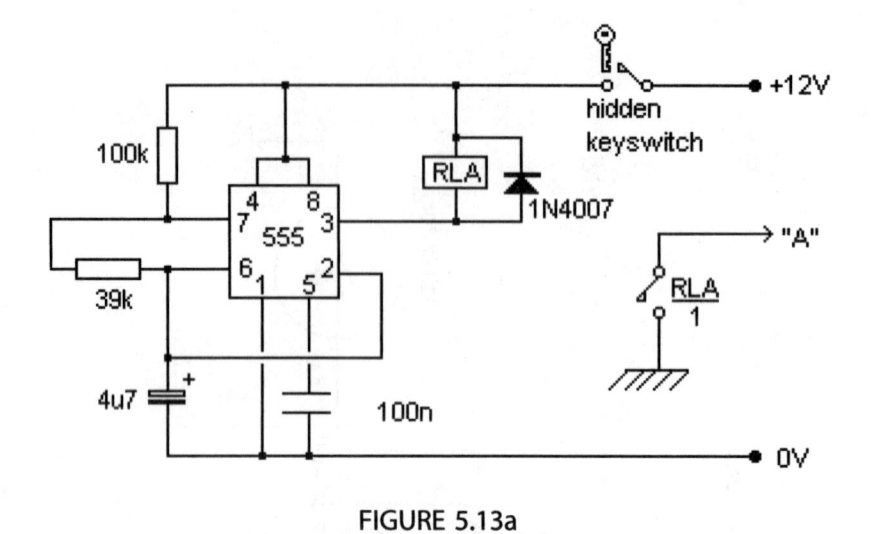

night, when compared to this "faulty vehicle" deterrent idea. To avoid possible detection of the device being caused by a loud-clicking relay, the chosen relay must be either a quiet device or be encapsulated in some form of noise insulation material.

Thermostat/Ice/Temperature Alarm

Figure 5.14 shows a simple thermostat, or frost/ice, alarm based on an LF351 op-amp IC. When commissioned as an ice/frost alarm, the sensitivity or calibration of the circuit is performed by the 1k variable resistor, which, once set, correctly should be sealed with an appropriate glue to prevent vehicle vibrations causing any problem. The negative temperature coefficient thermistor should be mounted externally, away from any heat source, and preferably in a shielded area to prevent the windchill factor, caused by the vehicle's traveling along, from giving a misleading indication. The thermistor and its flying leads must be made absolutely watertight, if possible. Calibration can be achieved by using an accurate thermometer and a glass of water with a large measure of ice cubes, noting that water begins to freeze at 4 degrees centigrade. Circuit operation is straightforward. The thermistor device has a

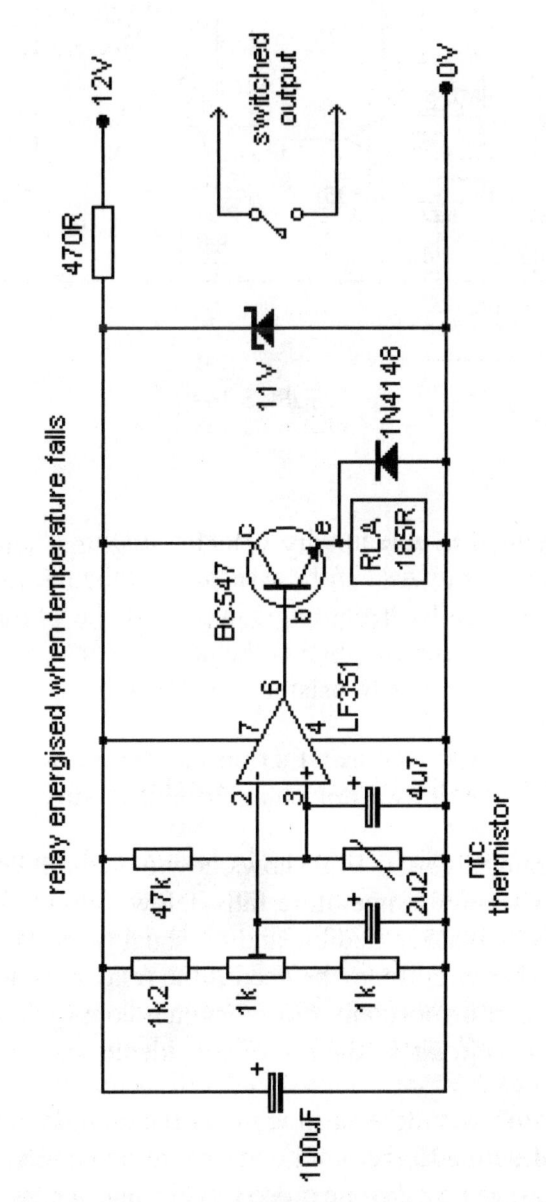

FIGURE 5.14

APPLIED SECURITY DEVICES AND CIRCUITS

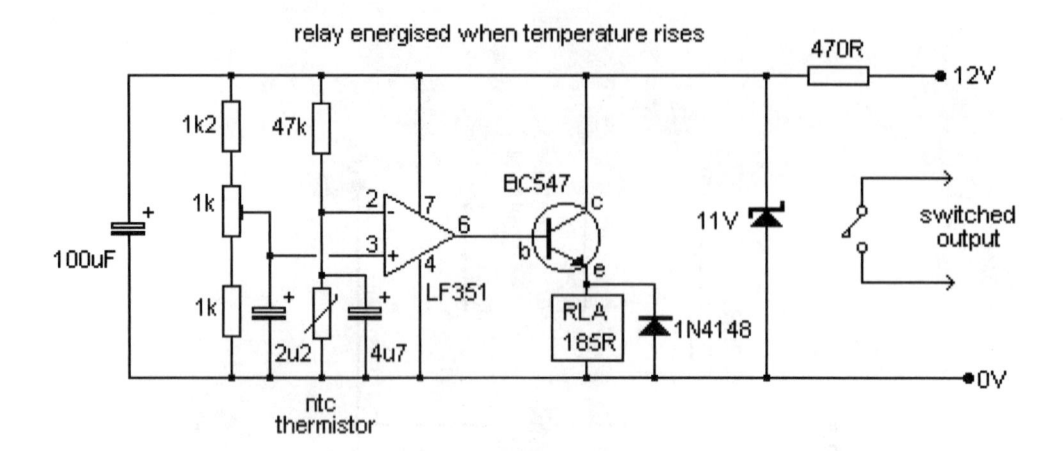

FIGURE 5.14a

negative temperature coefficient; when the temperature of the device rises, the resistance of the device lowers. If the resistance of the thermistor lowers, then the voltage on pin 3 falls, becoming lower than the preset voltage on the inverting input, pin 2. Because pin 3 is the noninverting input, the output also falls, so that the relay driver transistor switches off.

If the circuit is commissioned as a thermostat, the relay contacts feed power to a heater, with the thermistor device in close proximity to the heated space.

The second diagram, Figure 5.14a, shows how the circuit can be modified to activate a relay when the temperature falls below a preset level, so the relay contacts then may activate an audible and/or visual warning device, indicating ice conditions. Either circuit may be used if the relay available has the appropriate normally open or normally closed switch contacts, so long as current consumption by the circuit is taken into consideration.

Care must be taken when driving in icy conditions, because although the ice warning device described here is a good indication of conditions, there may be areas of roads, such as those exposed to chilling winds, with unexpected sheets of ice on an otherwise clear road.

Vehicle Tracking Devices

THE FOLLOWING THREE CIRCUITS USE RADIO-FREQUENCY TRANSMISSION AS A COMMUNICATION MEDIUM. NOTE THAT IT MAY BE ILLEGAL TO BUILD, USE, OR EVEN SIMPLY POSSESS THESE DEVICES IN YOUR STATE OR LOCALITY. CHECK FIRST!!!!

Vehicle tracking devices need no introduction. They sometimes are used to track the whereabouts of a suspect vehicle, or possible victim's vehicle, by various security forces. Tracking devices also may be hidden inside containers or boxes at risk of being stolen, so that security quickly can scan a warehouse to see if the goods have been hidden there and save precious time having to open up and unpack containers.

On a much lighter note: Tracking transmitters can be used as proximity annunciators, informing someone monitoring the frequency that you are only a mile or so from home, so it is time to put dinner in the microwave, for example. These transmitters also may be used for direction-finding games, where the transmitter is hidden up a tree or wherever, and the hunting parties must track down and pinpoint the device within a set time.

Three circuits for tracking transmitters are shown in Figure 5.15, Figure 5.16, and Figure 5.17. The first tracking device, seen in Figure 5.15, is the most economical to build, using a 555 audio oscillator whose output, a continuous stream of quick beeps, is fed into a simple VFO, or variable-frequency oscillator. The frequency on which this VFO transmits can be set by the constructor and is controlled by the LC combination in the collector circuit of the transistor. Because both of these components affect the frequency, they both may be made variable, but because the ferrite sores found in ferrite-cored inductors have a nasty habit of disintegrating during "tune-up," I prefer using trimmer capacitors. The coil L is made by winding five turns of thin, enameled copper wire around a 6mm form, such as a pencil, for example, and then slipped off the form after the enamel has been carefully scraped off with a craft knife. With a coil thus fabricated, when connected in parallel with a 4.7pF capacitor, the output frequency will be around 100MHz.

Automobile "tracking" transmitter using 555 ic with VFO transmitting section

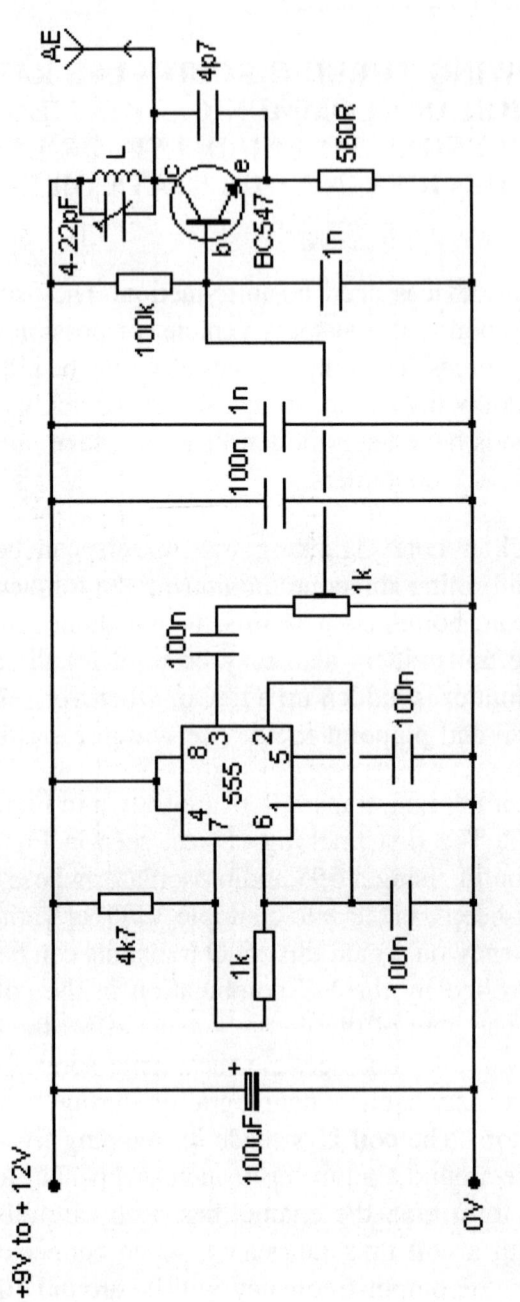

FIGURE 5.15

AUTOMOBILE SECURITY 83

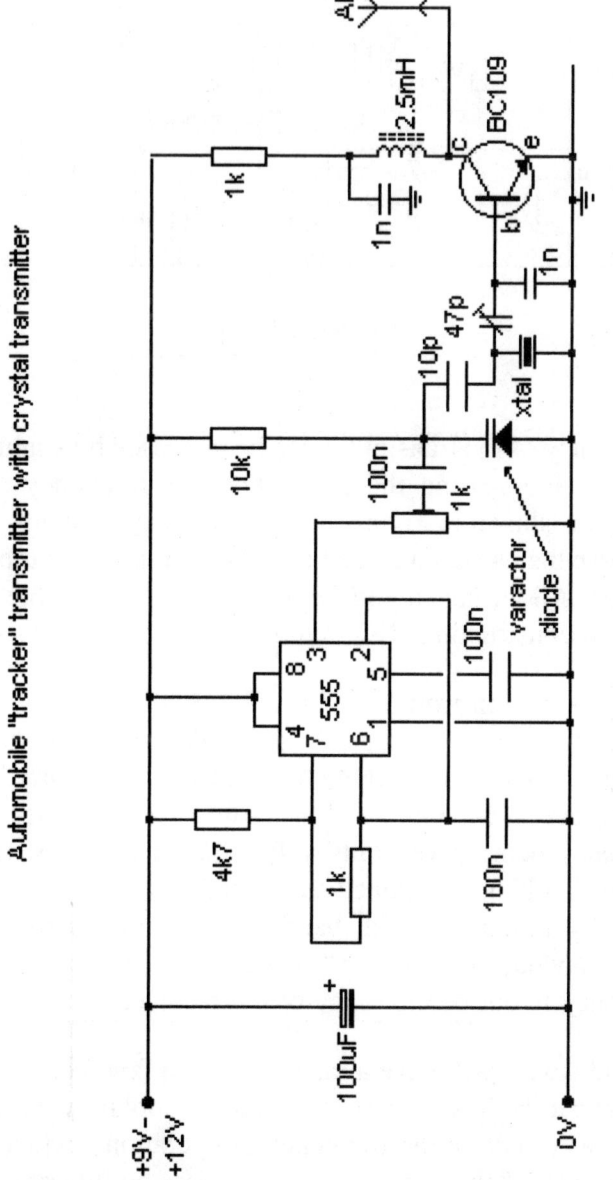

FIGURE 5.16

Automobile "tracker" transmitter with crystal transmitter

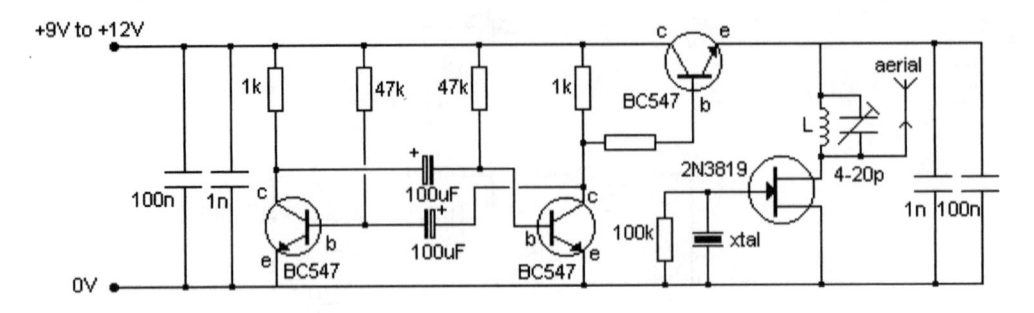

Automobile "tracker" with crystal transmitter, on/off carrier

FIGURE 5.17

The output frequency of the transmitter can be increased by slightly stretching the coil, whereas compressing the coil lowers the frequency. If a trimmer capacitor is used (as shown in Figure 5.17), closing, or meshing, the capacitor vanes will increase the capacitance and thereby decrease the output frequency to around 80MHz. All wiring should be kept as short as possible, as is always the practice when constructing RF circuits.

The amount of modulation from the 555 audio oscillator is controlled by the variable resistor and is more than adequate. Overmodulation will only make the already impure output from this circuit even more "dirty," with many signals occurring over a wide range of the radio-frequency spectrum. To find the main frequency being given out by this transmitter, a VHF (very high frequency) radio should be positioned about 50 feet away from the transmitter and then carefully tuned over the band. If the radio is too close to the transmitter, you may tune to a weaker "side-tone" frequency, not realizing that the main frequency is out of the range of the receiver!

These weaker side tones will have a range of only a few feet before they fade away. The aerial can be a foot or so of insulated hookup wire, the length of which will have an effect on the frequency of operation—which brings us to the biggest drawback of this type of circuit. Unless built very well, the VFO is notoriously unstable, with the frequency tending to drift all over the place if a hand moves near the aerial, circuit, etc.

The circuit may be powered either from the automobile's own battery or, if required, from any 9V to 12V battery. If the circuit is meant to be used in a clandestine fashion, any radio in the target vehicle would have to be rendered unusable; otherwise, the radio possibly would pick up the beeping tones produced from this or any other similar device!

The circuit shown in Figure 5.16 is somewhat similar to the preceding one, but instead of having a VFO transmitter section, it has a crystal-controlled oscillator. The crystal-controlled oscillator is very stable but has the disadvantage of being rock-bound, that is, the transmitter cannot move far from the frequency set by the crystal being used. If a high frequency is required from the transmitter, then frequency multiplier circuits must be used; these circuits increase cost and complexity in assembly and tuning up, plus they are physically larger circuits. The circuit shown does not have a tuned circuit in the collector circuit of the transistor, because the device is meant only as an example, requiring no setting-up equipment, and will operate with a wide range of crystal frequencies. Ideally, there would be an LC circuit here once a particular crystal frequency is selected. The antenna length in this design is not critical; the best length depends on frequency of operation, being longer for lower frequencies.

Although a varactor (varicap) diode is specified for the successful frequency modulation of the crystal oscillator section, you may do well by experimenting with a standard silicon diode, such as the 1N400x series. Although not widely known, these devices do display a measure of varying capacitance in their PN junction, and although it may vary only slightly—perhaps only half a picofarad or so—it sometimes is just enough to give some modulation in this type of circuit!

The frequency of this type of transmitter may be altered slightly from the crystal specification by adjusting the trimming capacitor that is in series with the crystal.

Unlike the first two examples, the tracking transmitter shown in Figure 5.17 will not transmit a series of audible beeps to the receiver but will transmit only an unmodulated carrier wave that is switched on and off by the multivibrator circuit, which keys the power to the crystal-controlled oscillator. This means that the receiver must be fitted with a BFO, or beat-frequency oscillator, in order to resolve any signal. Because there is no audio signal being transmitted, any radio scan will pick up a silent signal, rather than a giveaway beeping signal.

It is possible to build a keying circuit from a 555 timer IC that has a very unequal on-off ratio, for example, 30 seconds on and 10 minutes off, so that the tracking transmitter, hidden inside a container or such, transmits only for this short period of time, therefore narrowing the chances of being scanned by the target.

Automobile Infrared Tracker Beacon

If it is necessary to keep a vehicle under observation at night, security forces occasionally mark the vehicle by attaching an infrared-light beacon to that vehicle. Figure 5.18 shows a circuit that produces flashes of invisible infrared. This circuit uses a 555 timer IC in an astable mode and, instead of a visible LED for the output, it uses an infrared LED. Although only one LED is shown connected to the output, other LEDs may be connected in parallel, so long as

Automobile infrared "tracker"

FIGURE 5.18

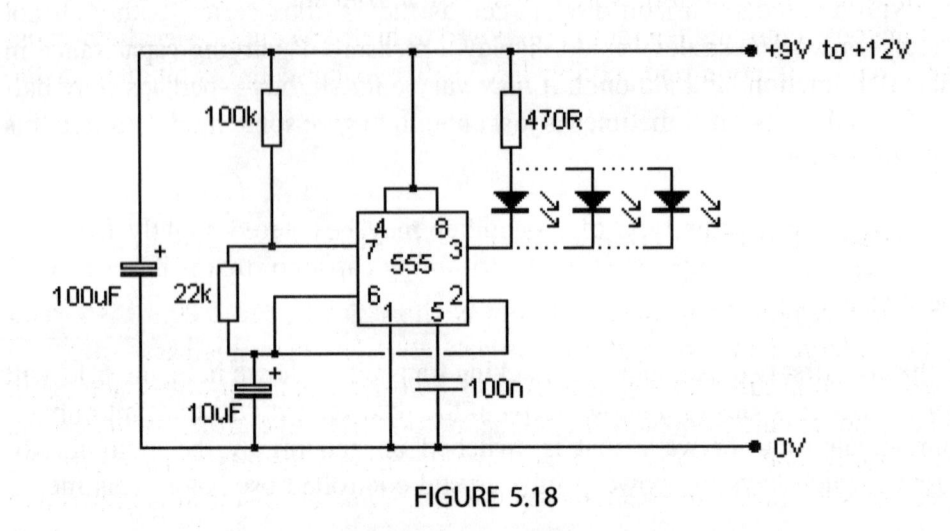

the maximum sink current of the device—200mA—is not exceeded. In this particular design, the 470R resistor is included to act as a current-limiting device. If a larger current load is required, the output of the 555 IC can be connected to a transistor, as shown in similar circuits in this book. Although security forces would tend to use an infrared night-vision device to look for infrared tracker beacons, an experimenter may want to try a standard CCD video camera, which may be sensitive to this wavelength and somewhat more easily available.

Automobile Radio Theft Alarm

The circuit shown in Figure 5.19 is designed to be activated if a radio is removed from a vehicle but also can be employed in other applications. It works by monitoring the grounding/chassis connection of the equipment being monitored. This connection is at ground potential until broken, which then, by the processing action of the three NOR gates, produces a negative pulse on pin 2 of the timer IC. This makes the timer go through a cycle, timed by the RC network, for about one minute before going back to an off state. During this on time, the relay is energized, thereby activating the chosen alarm sounder device. The normally open reset switch may be omitted or, in case of a fault in the circuit, replaced by a hidden on/off switch. The chosen sounder may be a high-powered siren installed inside the protected area, so that anyone removing the radio is exposed to high-level noise—perhaps causing a panic situation and forcing him or her to leave the equipment at the scene.

Automobile Hazard/Alarm Lamp Flasher

A simple vehicle lamp flasher design is shown in Figure 5.20. This circuit switches large lamps on and off automatically. The circuit is based on a 555 timer IC configured as an oscillator to give an equal mark-to-space ratio. When the circuit is switched on, the output from the timer switches the transistor on and off for an equal period, thus turning the relay on and off. The speed of switching can be altered by the 1M variable resistor. It is imperative

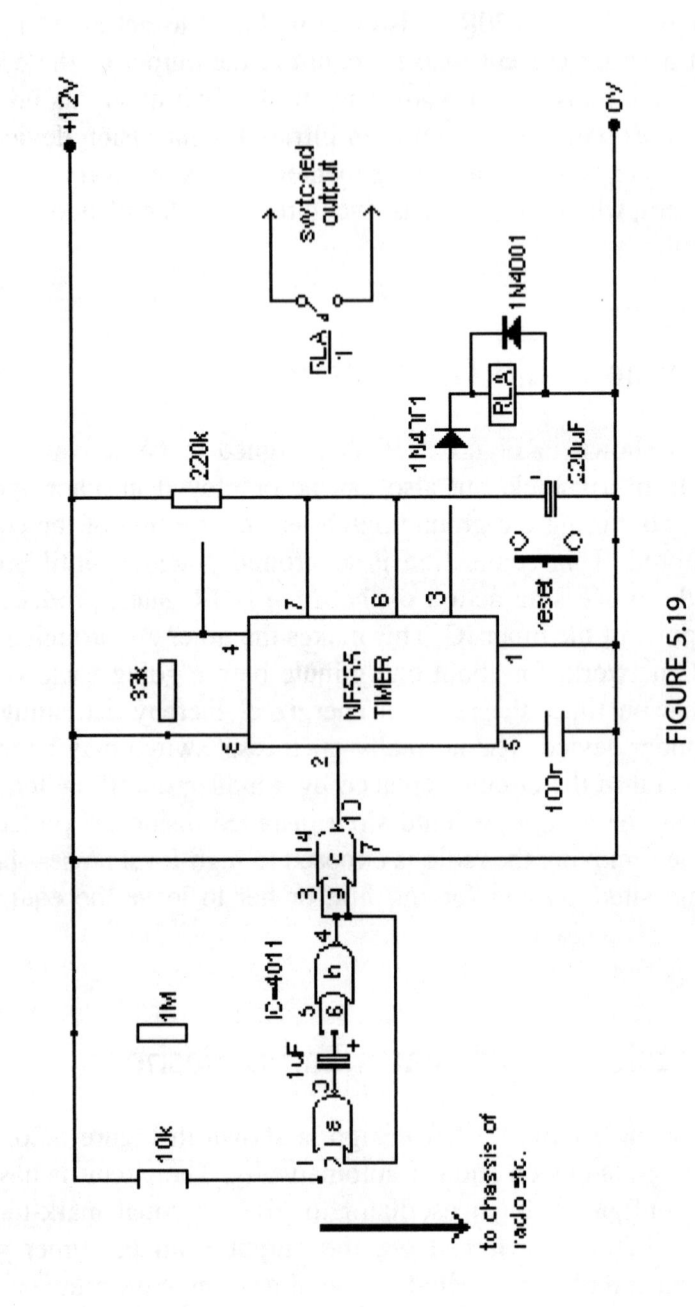

FIGURE 5.19 Automobile radio theft alarm

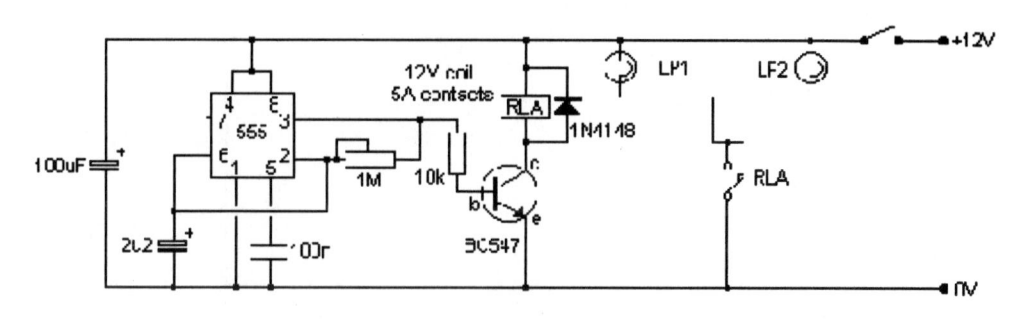

FIGURE 5.20

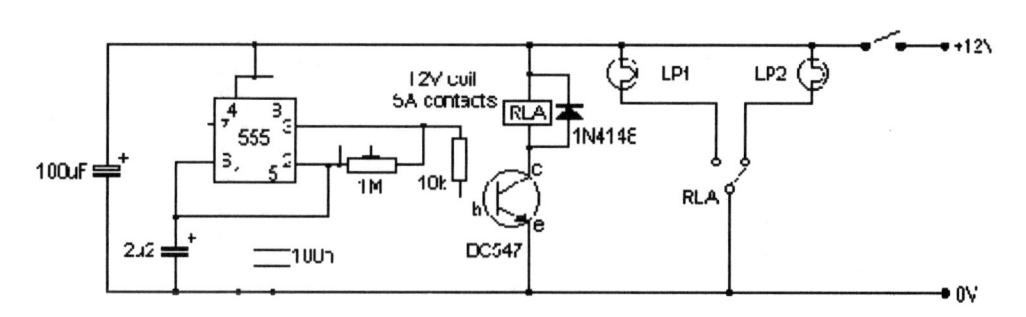

FIGURE 5.20a

that the current rating of the relay contacts are not exceeded, or they may either burn out or weld themselves together, especially if high-powered lamps are used or if several lamps are connected in series. Because a point during switching the circuit off possibly will occur when the lamps are on, the power switch also must be suitably rated.

An alternative arrangement is shown in Figure 5.20a. This circuit is almost identical to the preceding design, with the difference being that the relay used has an extra pole. When the relay is not energized, LP1 is illuminated, and when the relay switches over, LP1 is extinguished and LP2 illuminates. This gives an alternating lamp display similar to those eye-catching displays used by some security forces.

CHAPTER 6

Operational Amplifier Circuits

The op-amp is useful for circuits that need to give an output if certain conditions on the input pins are altered only minutely. Stated briefly: The device has two inputs, one called the *inverted input* and marked up on diagrams with a minus sign and the other called the *noninverting input* and marked up with a plus sign. If you are unfamiliar with these devices, do not confuse the inputs and the plus and minus signs of the supply!

The following are some important things to note when designing op-amp circuits for the first time:

▲ A positive applied to the inverting input makes the output go down, that is, the input signal is inverted by the amplifier.

▲ A positive signal applied to the noninverting input makes the output go up.

▲ The gain, or amplification, of the device is controlled by applying feedback from output to input.

Note that the positive signal has to be only a bit more positive than the negative signal in some op-amp configurations.

The 741 op-amp is shown in Figure 6.1.

Operational amplifier-
the 741

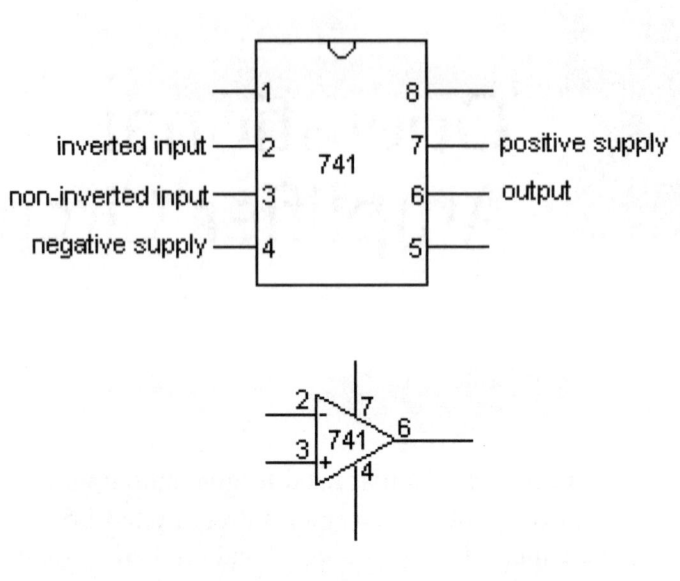

- 1
- inverted input — 2
- non-inverted input — 3
- negative supply — 4
- 8
- 7 — positive supply
- 6 — output
- 5

FIGURE 6.1

CIRCUITS USING OP-AMPS

A multitude of circuits use the op-amp device, and because operational amplifiers lend themselves to amplifying the relatively small changes produced by sensors such as light-dependant resistors, thermistors, and power supply hum, the op-amp device proves very useful.

Over-Temperature Alarm with Relay Output

Figure 6.2 shows a circuit whose op-amp output switches a transistor to operate a relay if the temperature sensed by the thermistor is increased, the value of which is set by the 10k preset resistor. This circuit could be used to monitor the freezer in a home. In the case of power failure, a battery backup system would be used to power the circuit and associated warning device.

OPERATIONAL AMPLIFIER CIRCUITS   93

Over-temperature alarm with relay output

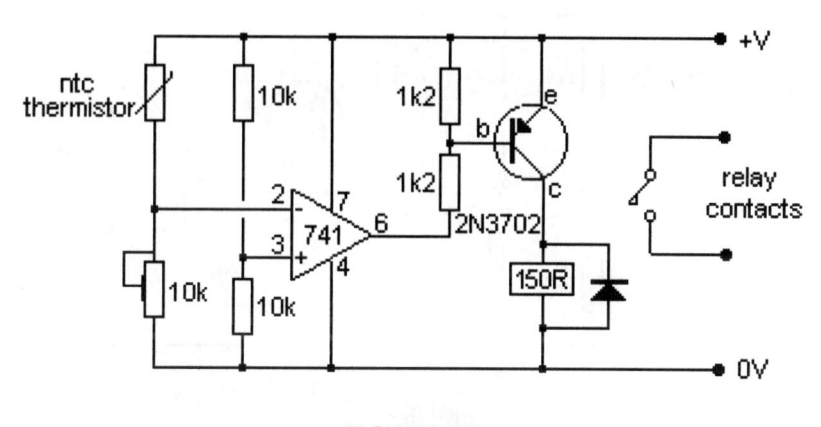

FIGURE 6.2

Frost / ice warning with relay output

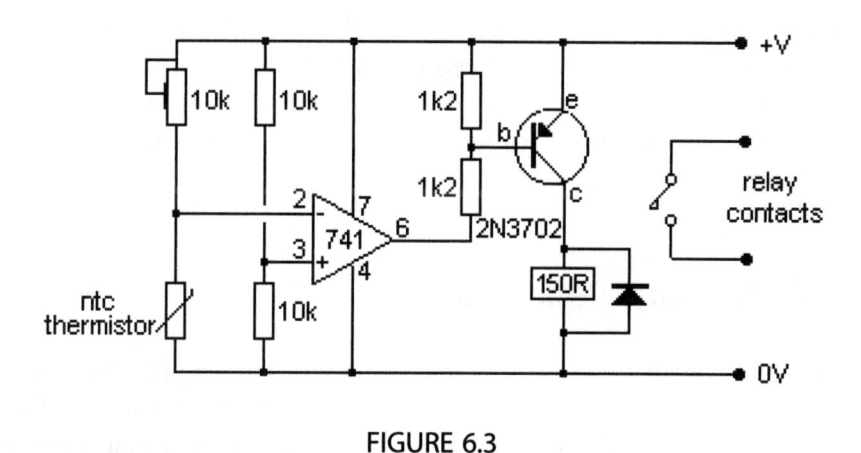

FIGURE 6.3

Frost/Ice Warning with Relay Output

Figure 6.3 shows that, by swapping over the preset resistor and thermistor, the output becomes active if the temperature falls below a preset level. If the circuit is to be used in an automobile, great care must be taken to ensure correct calibration, as well as the correct siting of the sensor.

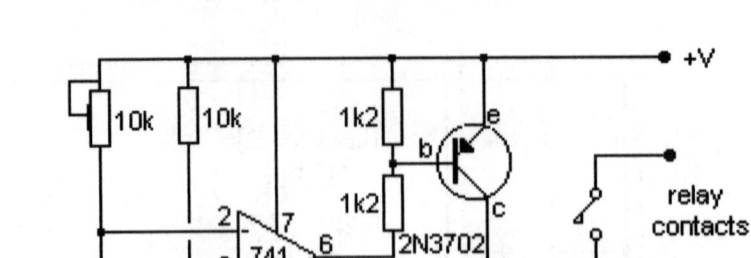

"Dark" alarm with relay output

FIGURE 6.4

Dark Alarm with Relay Output

Figure 6.4 is identical to Figure 6.3, except that the NTC (negative temperature coefficient) thermistor is replaced by an LDR (light-dependent resistor). This could be used to give an alarm if darkness has fallen, perhaps because security lights have failed, or to warn a farmer that it is time to secure the chicken house.

Light Alarm with Relay Output

The circuit shown in Figure 6.5 gives an output if the light level increases above the preset threshold, so it could be activated if security lights come on, or, if the aforementioned chicken farmer was fast asleep, it could trigger an alarm or open the door to the chicken house at the break of dawn, if connected to a suitable actuator.

Electronic Thermostat

Figure 6.6 shows a circuit for an electronic thermostat, which gives an accurate control to the nearest degree C, with a range between just above freezing

Light alarm with relay output

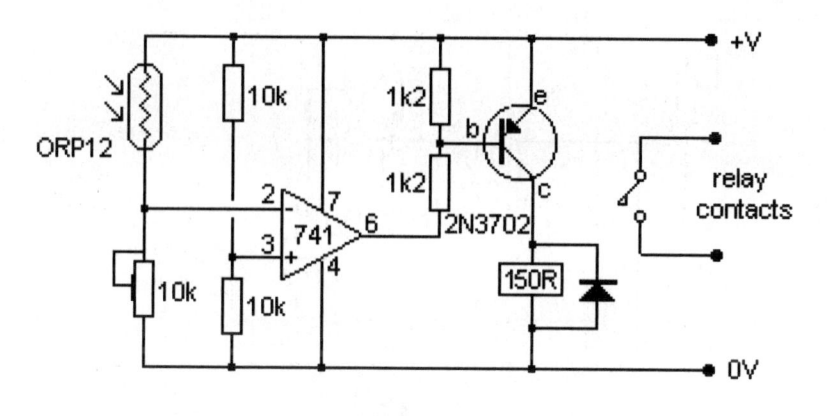

FIGURE 6.5

to around 30 degrees centigrade. The variable resistor acts as the temperature control setting, so it may be pertinent to fit it with a knob with a pointer. Because the unit is powered from the mains and will be controlling mains voltages, you must take great care to avoid electric shock. If the relay is to switch on large current loads, make sure that the contact ratings are sufficient. You may need to connect suitable high-voltage rating capacitors across the relay switching contacts to avoid excessive sparking causing RFI (radio-frequency interference) to nearby radio and television receivers. Also make sure that the cable that is to carry the load current is of sufficient capability. As with all thermostats, carefully consider the position of the sensor.

Touch/Hum Switch with Relay Output

Figure 6.7 shows a circuit that can be triggered by touching the two terminals simultaneously with a finger. If the circuit is used within an area containing a large amount of power supply hum, then it may be sufficient just to have one terminal, that is, the one going to the op-amp. If the gain of the selected op-amp is turned up even higher, the switch may be activated by placing the finger near to the terminal, which could be a small metallic plate fastened to the inside of a shop window so that a prospective customer could switch on internal displays such as lighting or automaton-type toys.

FIGURE 6.6

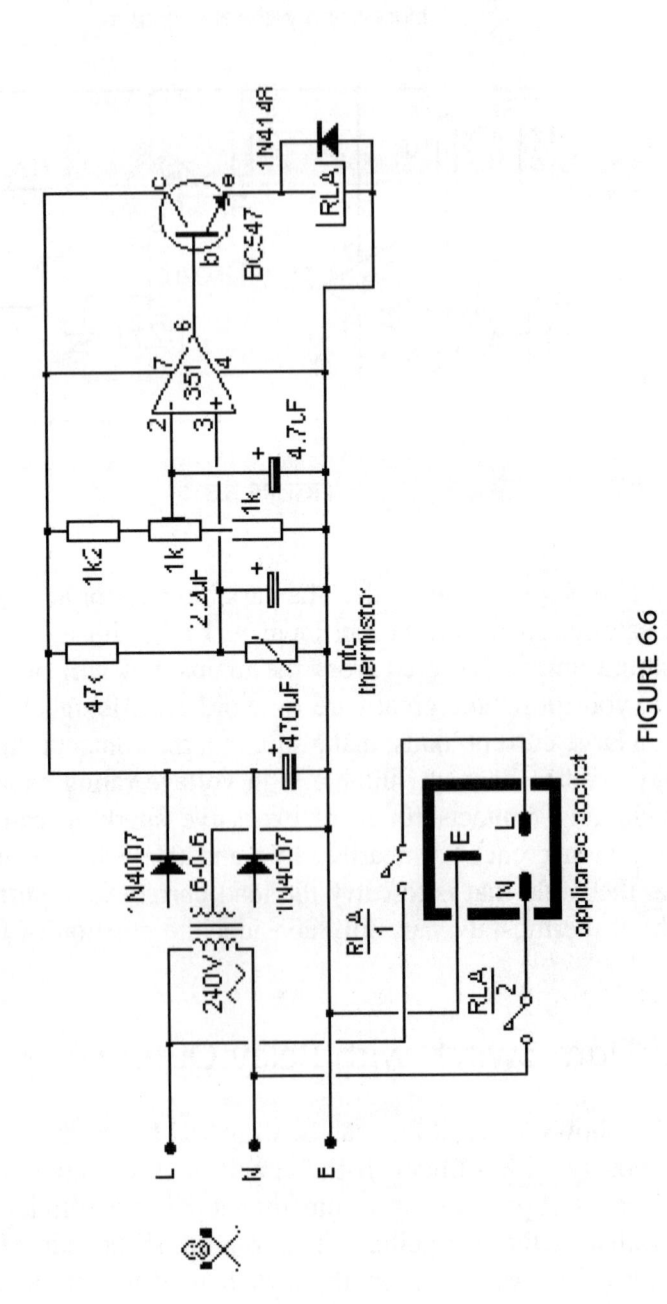

Touch / hum switch using the 741 ic

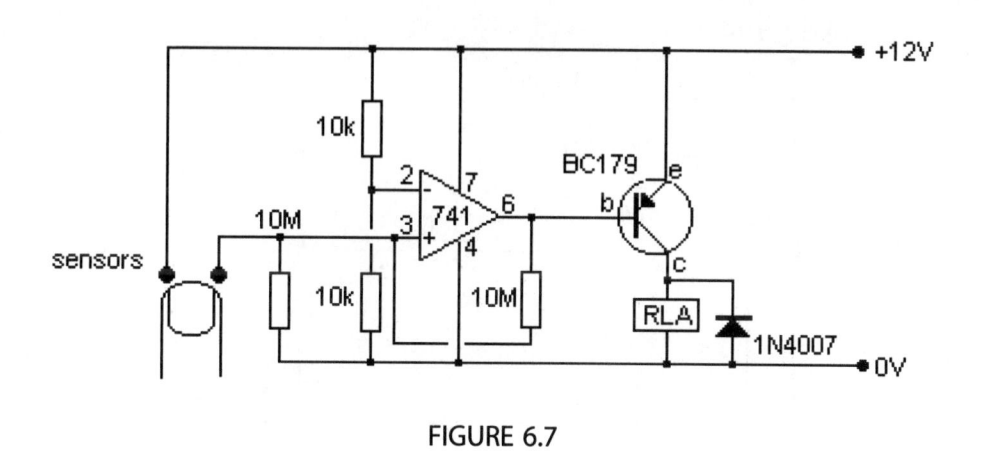

FIGURE 6.7

Battery Condition Monitor Using the 741 Op-Amp

Figure 6.8 shows a simple circuit that monitors the condition of a battery, such as the backup battery in an alarm system. One input leg is stabilized at the working specification of the zener diode—in this case, at 5.6V—whereas the noninverting input of the op-amp is connected to a resistor chain so that the sensitivity (turn-on or turn-off point) can be set by the user. By simply swapping over the circuit inputs, you can arrange for the LED to come on or off when the criteria have been reached, although, for the sake of battery consumption on a system not being recharged, an LED continuously drawing current from this source obviously is not advisable!

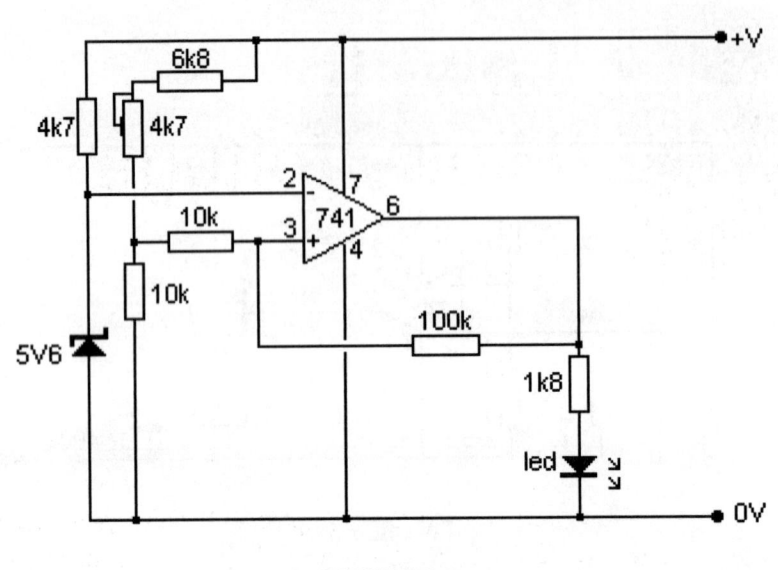

FIGURE 6.8

CHAPTER 7

Low- to High–Voltage Circuits

The three circuits presented first in this chapter will convert a low-voltage source of approximately 12V DC up to a high-voltage, low-impedance source of approximately 300V DC. An example of their use could be as a driving source for small electric fences to prevent chicken houses from attack by foxes or to prevent animals from crossing into adjacent areas of a field.

> **WARNING**
>
> These devices should never be used as "mantraps," and wherever used, prominent warning signs should be displayed, and/or whatever local laws dictate should be followed to the letter. Although a shock from these types of circuits may be just unpleasant to some, certain circumstances may present secondary complications if, for example, the recipient has a weak heart or falls into a road and gets run down by a truck.
>
> Because these designs have a capacitor connected in parallel across their outputs, remember to short the output connections together after turning the supply to the circuit off and before working on the

> unit, because these capacitors will remain charged at the high voltage for a considerable amount of time!
>
> Any electric fence requires two conductors to deliver the charge. This may be by using pairs of conductors—or sets of pairs of conductors, as shown in Figure 7.1—or by replacing one of these conductors with a comprehensive grounding electrode system, as shown in Figure 7.2.

The advantage of a one-conductor system is that installation costs can be lower; however, the disadvantage is that multiple grounding points may have to be employed. This is because, since the one-conductor system relies on the actual ground to act as the "missing" electrode, if the ground is not very conductive—for example, it has very low moisture content—then, to compensate for this, the voltage must be turned up higher so that the points further

Two-conductor method

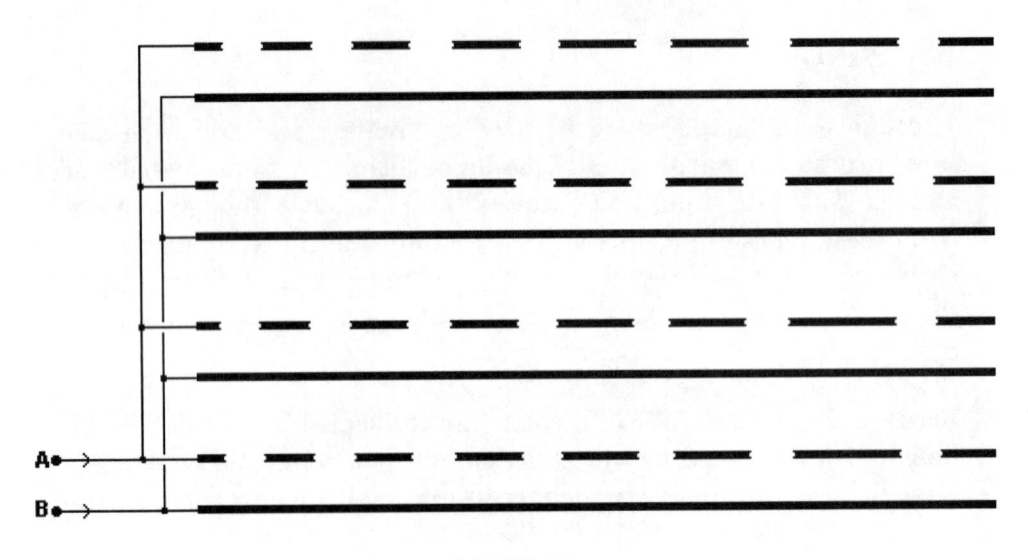

FIGURE 7.1

LOW- TO HIGH–VOLTAGE CIRCUITS

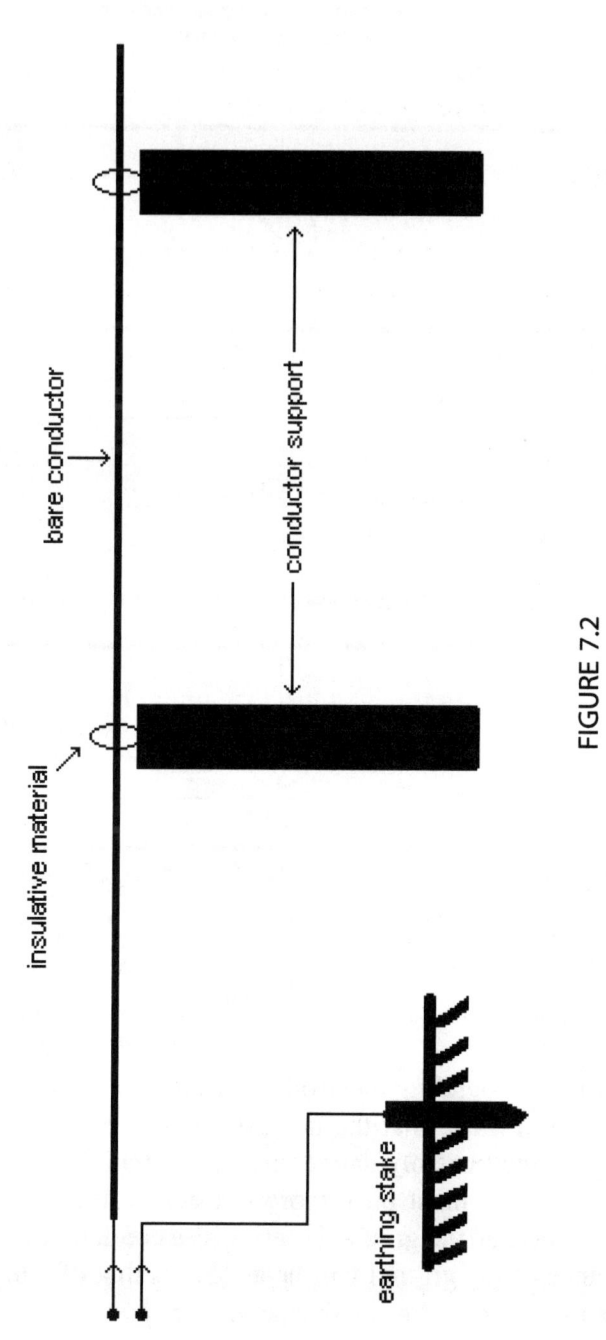

FIGURE 7.2

Problem caused with high-resistivity area when using one-conductor system

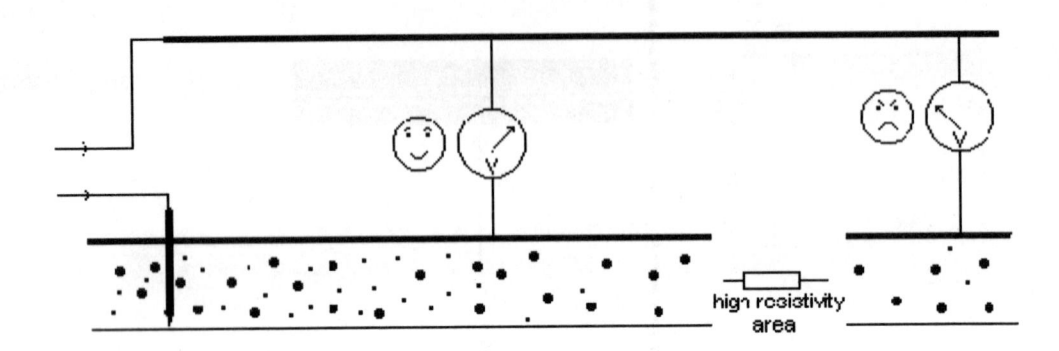

FIGURE 7.3

Using a "booster" electrode to overcome high resistivity of ground

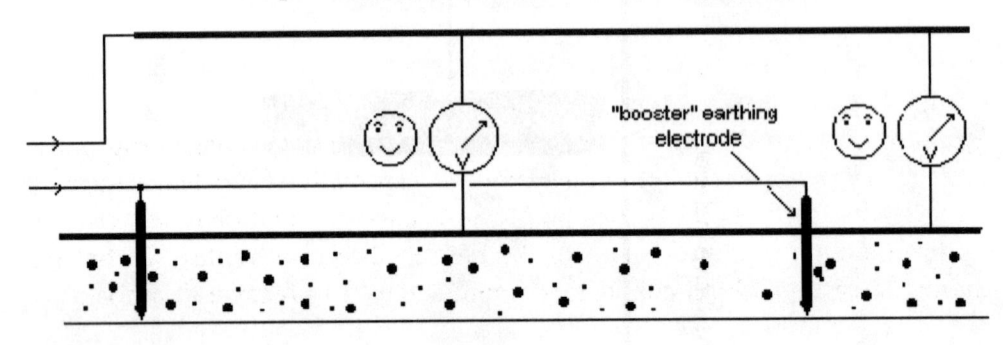

FIGURE 7.4

away from the grounding electrode still are as effective, as shown in Figure 7.3. This would mean that the nearest point to the grounding electrode is at a very high—and possibly dangerous—potential. To produce a balanced system, you therefore must bury more booster ground electrodes, as shown in Figure 7.4. Because the ground is acting as a conductor, there is no chance that the resistance of the ground will be as low as that of a length of copper wire; therefore, to overcome this inadequacy, more power is always required in one-conductor systems than in two-conductor systems.

High voltage generator using a relay or electromechanical buzzer

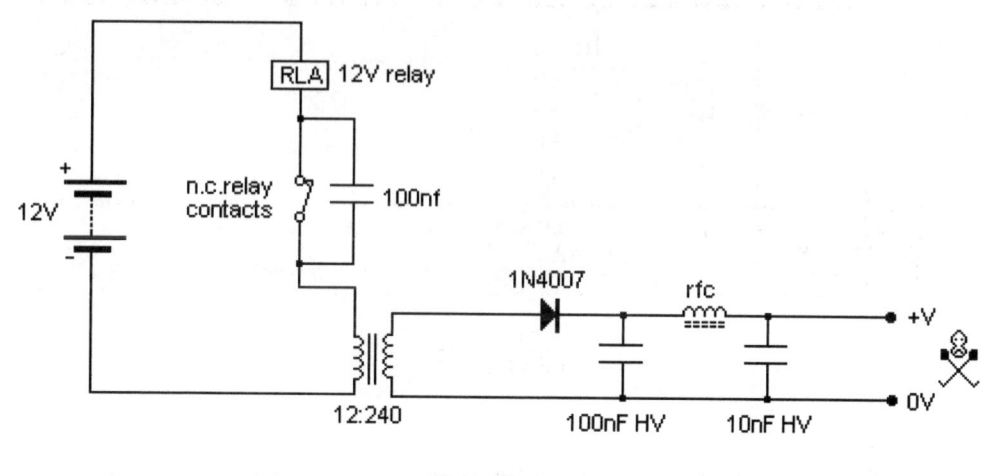

FIGURE 7.5

The advantage of a two-conductor system is that, if any voltage drop in the conductors is not taken into consideration (especially if the fence forms a circle so that ends of conductors meet), the voltage distribution is near enough uniform along the entire length of the barrier. The disadvantage is that the animal must touch both conductors simultaneously to receive the shock.

The most simple of circuits to convert low voltage to high voltage is obtained by using an electromechanical buzzer or relay, as shown in Figure 7.5. The design works on the same principle as old military tube radios did to obtain the high voltage required for anode voltages and the like. The low voltage is fed to a relay coil, which is connected in series to a pair of its own normally closed contacts, and then to the low-ratio side of a transformer. Because the circuit is complete, the coil activates the switch, thereby opening it up. Since there now is no power to the coil to keep it activated, the relay resets, closing the contacts. While the circuit is continuously opening and closing itself, the current in the transformer also is being switched on and off. This switching in the transformer primary is transferred to the secondary of

High voltage supply using centre-tapped primary transformer

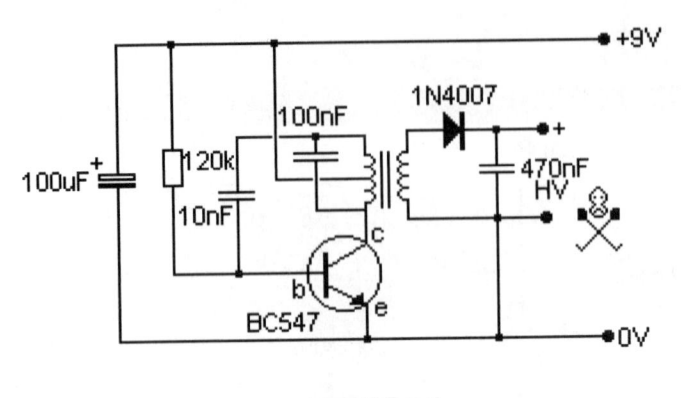

FIGURE 7.6

the transformer—the high-ratio side—so large voltages are induced, which then are rectified, and then energy is stored by the associated circuitry.

If a buzzer is used in place of the relay, the buzzer may be mechanically adjusted for maximum output, which could well exceed 1000V. The HV (high-velocity) capacitors should be suitably rated, but if the capacitance is increased in these circuits, small animals will not be deterred but most likely will be spot-welded!

A simple solid-state circuit is shown in Figure 7.6. This uses a single transistor in a Hartley oscillator configuration. The frequency of operation may be altered by replacing the 120k resistor with a variable resistor. As is the case in all inverter circuits, the transformer is used "back to front," as it were, with the low-voltage side being connected to the driver circuit and the high-voltage side now becoming the output. The impedance of this device is very low and, although producing an unpleasant jolt, is relatively safe. The transformer used is a 9V-0-9V center-tapped mains transformer, although a 6V-0-6V may be used if a higher voltage output is required. The low-voltage supply can be adjusted to suit both the transformer being used and output requirements, so long as the HV capacitor has a sufficient voltage rating.

High voltage generator using the 555

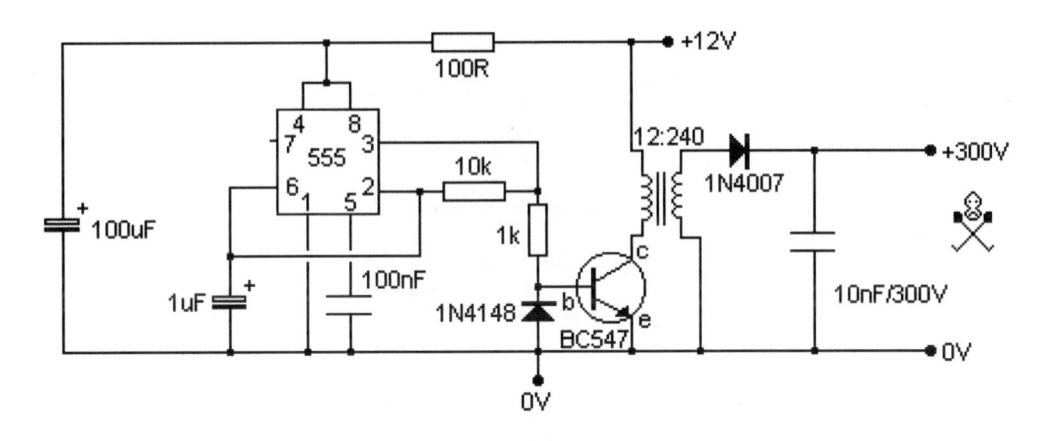

FIGURE 7.7

Another solid-state design is shown in Figure 7.7. In this design, the 555 IC is configured as an equal mark-to-space astable multivibrator, operating on a frequency of around 70Hz, which is close enough to the usual mains supply frequency of 50Hz or 60Hz, depending on the country. The output of the timer is fed to a current amplifier transistor, which switches the transformer on and off at the same frequency of the oscillator. The high-voltage AC (alternating current) obtained from the output of the transformer is then fed through a full-wave bridge rectifier to convert it into DC, and then power is stored in the high-voltage capacitor.

The last design, shown in Figure 7.8, is not an inverter in the same fashion as the three previous designs. It is powered from the mains supply but has an isolating transformer plus a 47k current-limiting resistor as safety devices. The circuit utilizes a thyristor that is triggered with the help of a gas discharge device, and then the power from the 1uF capacitor in series with the primary of the ignition coil is dumped into the ignition coil. Several thousand volts then are produced in the secondary winding of the automobile ignition coil,

106 APPLIED SECURITY DEVICES AND CIRCUITS

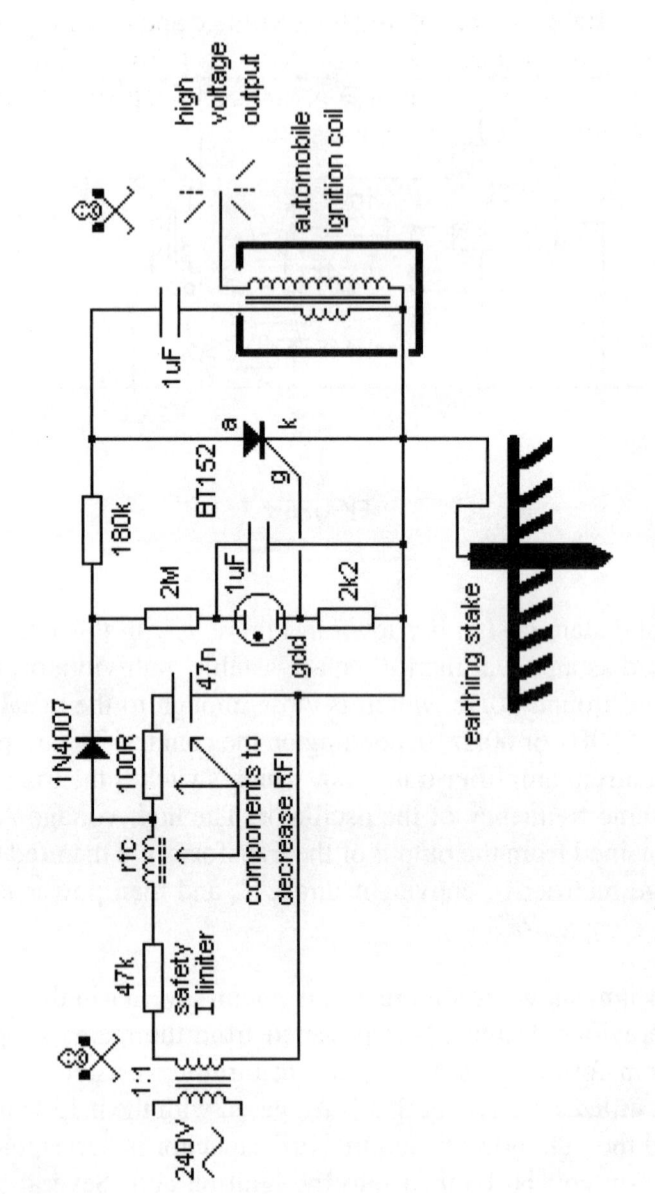

FIGURE 7.8 High voltage / low current source for electric fence

therefore this design is suitable only for the extremely competent and not-easily-shaken constructor! The circuit should be very well fuse-protected and kept dry at all times to prevent the high-voltage and extra-high-voltage points from arcing across to other points of the circuit. Some of the capacitors and GDD (gas discharge device) probably can be salvaged from the starter plug-in module from a fluorescent lighting unit.

CHAPTER 8

Miscellaneous Designs

The following miscellaneous designs are presented in a modular approach, where you can select some of the ideas to build a security system to suit yourself. Also in this chapter are many designs that will stimulate the designer into building individual modules for use as practical stand-alone units.

ANTI-TAMPER SWITCH PROTECTION

Although use of tamper switches is discussed in previous chapters, I think it's prudent to note how they are used in practice. If a circuit—which may be a bell box, keypad, intercom, or so on—is in an unguarded place, sooner or later it will come under attack. To warn that a vulnerable unit is under attack, tamper switches, due to their inherent robustness and low cost, provide an excellent choice.

Figure 8.1 shows how a cabinet is protected by fitting a tamper switch in such a way that, when the door is closed, the switch arm is depressed continually, but if the door is opened, the switch then springs open and so triggers an alarm. Note that the sensor arm of the tamper switch is bent by the installer so that physical adjustments can be made to the fitting, as well as making the switch more sensitive to the door of the cabinet's being opened, albeit minutely. The switch may be mounted on the side wall of the cabinet as shown, since if screws or other fittings are removed by an attacker, the tamper switch will drop away triggering the switch. A more professional look would be to mount the tamper switch on the back of the cabinet, making it invisible, on some type of positioning bracket, but this makes positioning the switch much harder, although no screws will be visibile to provide a temptation to a habitual meddler.

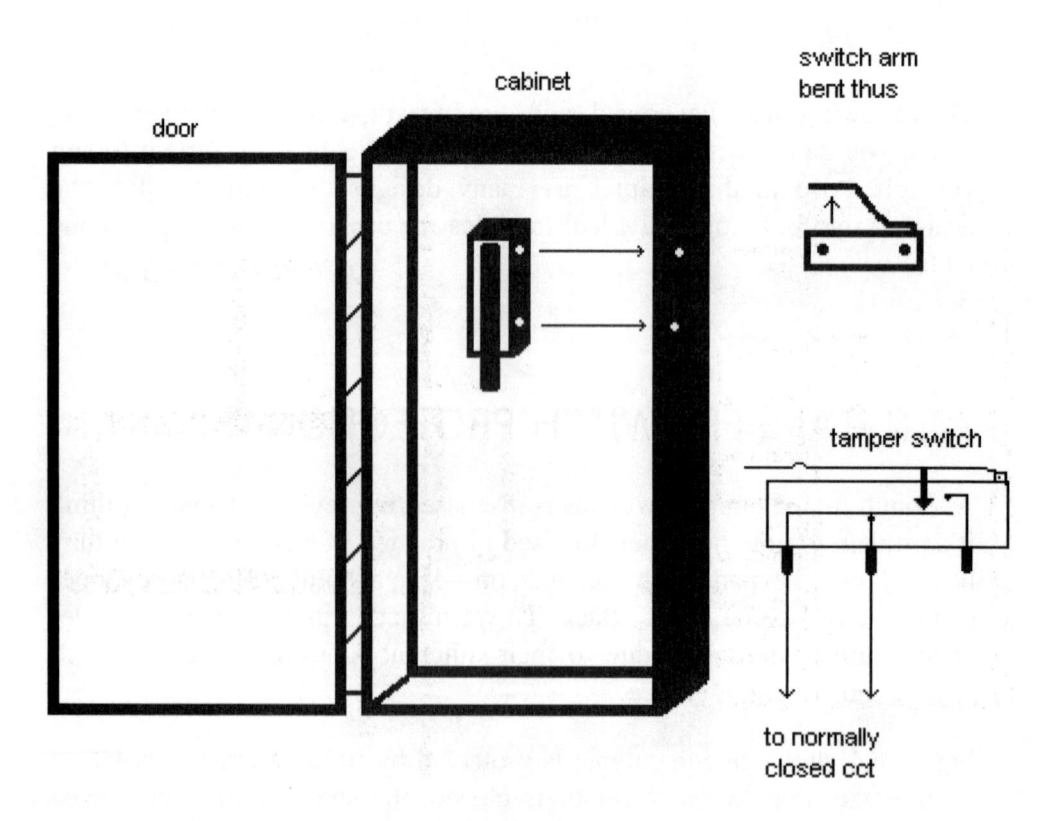

FIGURE 8.1

The next method of mounting a tamper switch is to give an alarm if the box or cabinet is ripped off the wall that it is screwed to. To this end, a tamper switch can be mounted on the internal side of the box as before, but this time the arm projects through the back of the box through a precut slot, big enough so as not to foul any movement of the arm, as shown in Figure 8.1a. When the box is screwed to the wall, the switch is depressed, but if the box is removed from the wall, the switch is triggered by the fact that the arm now is free. The problem here is that it is sometimes possible for the tamper arm to be held down by a shim of thin metal while the box is unscrewed or forced from the wall face. This becomes less of a problem if the switch is mounted on a free-pivoting bracket near the center of the box, as opposed to near the edge.

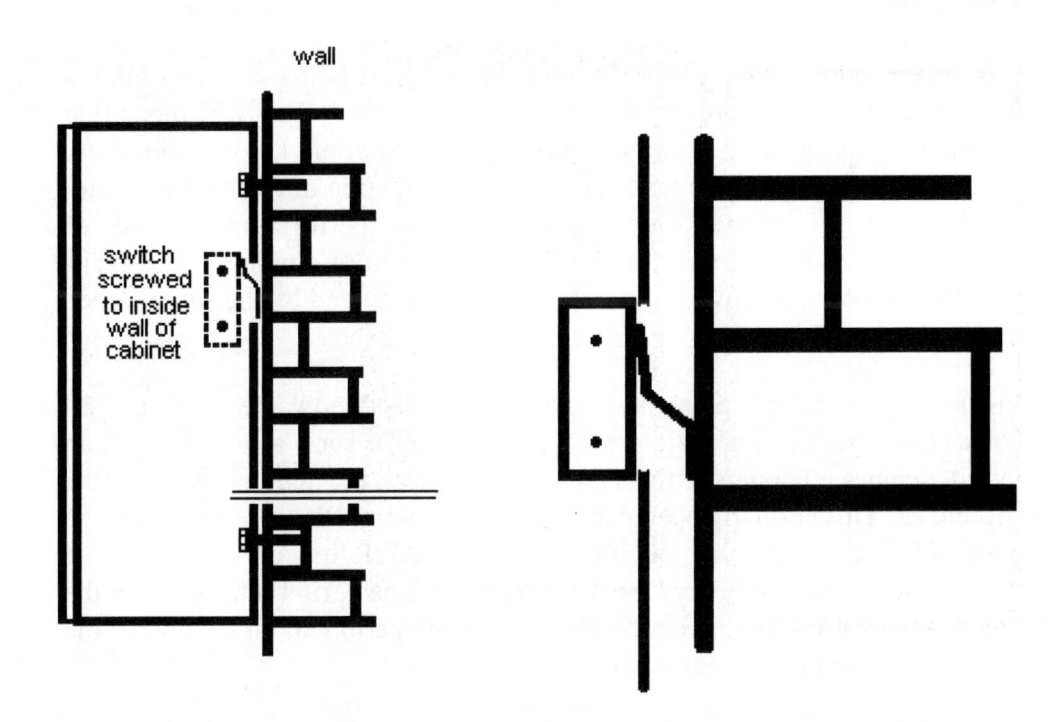

FIGURE 8.1a

The concept of generating an alarm if a bell box is ripped from a wall is more useful if the bell box contains a backup battery and circuit that will self-latch and automatically power the siren and strobe, so as to deter the attacker from walking down the road with your box!

Note that the illustrated anti–tamper switches throughout this book may have different pinouts to those that may be available, therefore the installer must test his or her own particular switches for correct operation by using a multimeter.

AUTOMATIC SURVEILLANCE CAMERA VCR SWITCHING

It is very useful to have some method or another of being able to switch a video recorder on if a protected area has been invaded. This will mean that the video recorder operates only if there has been some kind of violation, which means that valuable tape space, and the viewing of the tape at a later date, is then minimized. The easiest method of performing the task of switching the video recorder on is to have a system of commercially available PIR sensors that cover the same area as that covered by the surveillance camera.

Figure 8.2 shows a typical basic arrangement and how the units are all connected together. The PIR sensor is mounted in such a way that, if an intruder comes into range of the surveillance camera, the internal switch of the PIR closes. This normally open pair of contacts controls the video recorder via a suitable VCR controller circuit. Because the PIR has an integral timer circuit, the contacts are operational for only the time period set, therefore the video recorder is switched off after the preset period of time, unless the protected area still is occupied by a detectable body.

A simplified diagram of a VCR control switching unit is shown in Figure 8.2a. A double pole relay is connected in such a way that, once the coil is energized

Auto Surveillance Camera VCR Switching

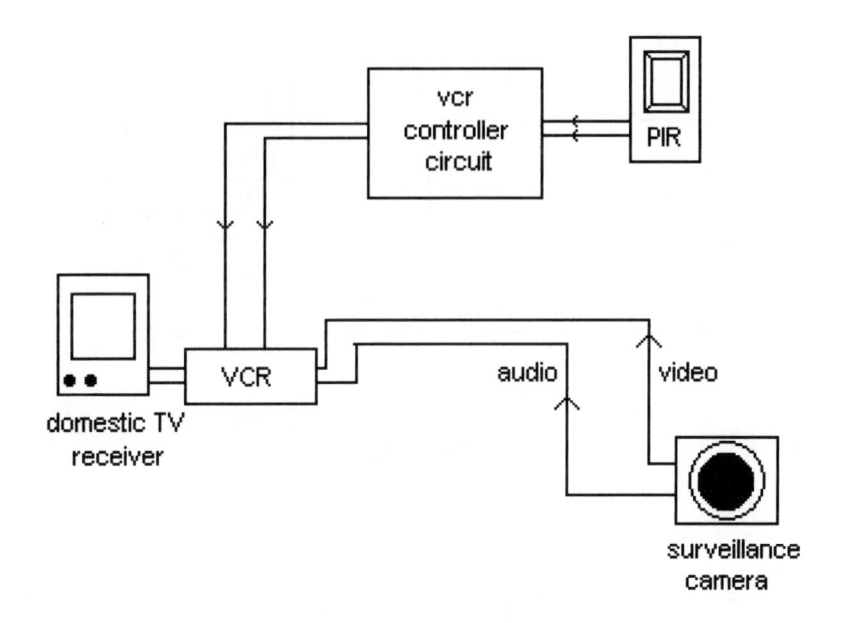

FIGURE 8.2

via the PIR internal switch and a suitable low-voltage supply, sets of contacts wired in parallel across the "record" switch of the VCR cause the recorder to go into the record mode. A further pair of relay contacts, which are normally open when the PIR relay and the controller relay are not powered up, will be connected in parallel with the "stop" switch of the VCR, thereby stopping the recording when the PIR is not triggered.

If a particular VCR model seems incompatible with the idea of having the record switch held down for any length of time, it is possible to use a more complex circuit that uses a 555 timer as a one-pulse device so that if a signal is received from the PIR, the 555 closes a relay for a second or two, so operating the recorder.

Auto VCR Switching Diagram

connected in parallel
with record switch
of VCR

connected in parallel
with stop switch of
VCR

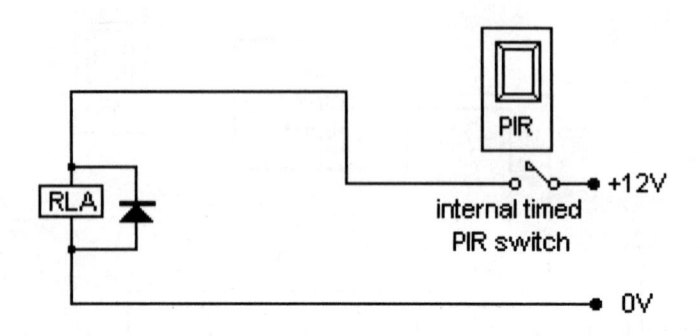

FIGURE 8.2a

AUTOMATIC LIGHTING CONTROLLED BY DOORBELL

Figure 8.3 illustrates a circuit that will switch on a VCR, floodlighting, or anything you wish for a preset length of time, if the switch is pressed. It will startle any prospective attacker if a 1kW floodlight illuminates his or her face

Automatic lighting switched by doorbell

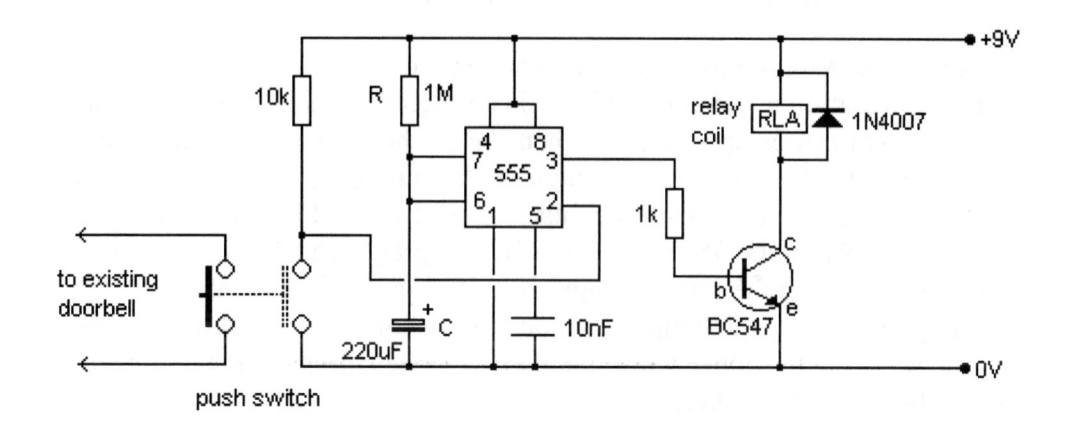

FIGURE 8.3

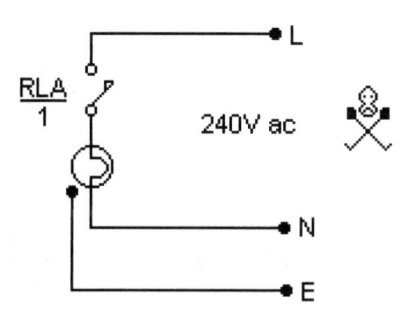

FIGURE 8.3a

as soon as they press the button! The device also may be used with a single-style pushbutton for lighting up alleyways, stairwells, and garages for a predetermined time in order to save on the electricity bill. The timed period for which the switch remains on is controlled by C and R, as covered in Chapter 3.

Figure 8.3a illustrates how the relay contacts are arranged to switch the power supply to the lamp. Note that the power supply must be switched off if you are servicing the unit, because if the bell button is pushed during servicing, the power supply will be switched by the relay contacts!

Figure 8.4 shows a minimum-component-count circuit for a light-timing circuit, also based on a 555 timer IC that is configured as an equal mark/space oscillator. On power-up—that is, when the pushbutton is pressed momentarily—the output of the 555 goes high, so latching the circuit on with relay contacts RLA1. The other pair of contacts now also are closed, thus switching on the lamp. When the 555 toggles low after a predetermined time, the relay switches off so that the lamp extinguishes. With a small modification to this circuit, you can have the lamp switch on after the predetermined time, so that anyone watching the house will think that someone still is in the building and hopefully will go elsewhere.

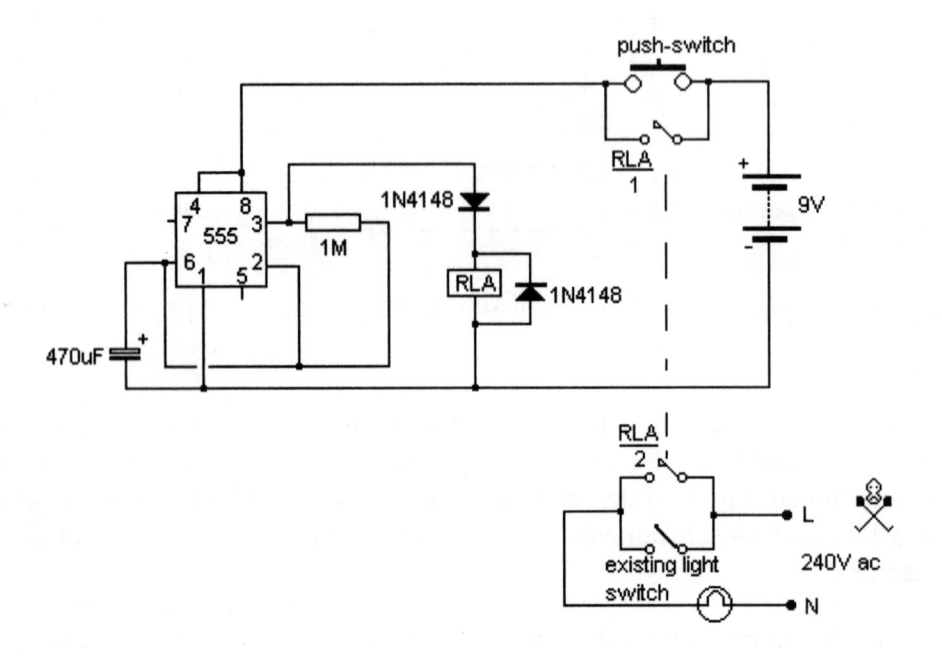

FIGURE 8.4

NOTE: AS WITH ALL RELAY CONTACT POWER-SWITCHING CIRCUITS, ENSURE THAT HIGH VOLTAGES ARE ISOLATED. MAKE SURE THAT THE RELAY CONTACTS HAVE SUFFICIENT CURRENT- AND VOLTAGE-HANDLING CAPABILITIES AND THAT RELAYS ARE OF SUFFICIENT FLASH-TEST RATING!!!!!!!!!

BATTERY BACKUP CHARGING CIRCUIT

A typical charging circuit for a 12V lead/acid backup battery-charging device is shown in Figure 8.5. The transformer, full-wave rectification diodes, smoothing capacitor, and voltage regulator are shared with the existing alarm circuitry. The additions for the charging circuit are the fuse, which will blow if the battery becomes a short circuit, and the diode Dp, which is a protection device. Imagine that a fully charged battery was connected in reverse—a simple mistake due to the spade-end terminals that are used quite often—to the alarm circuit. In this event, the diode conducts the reverse current through the fuse and back to the other terminal of the battery, thus blowing the fuse. The value of the fuse should be carefully chosen so that it will blow should any problem arise, but still be large enough to accept the large current taken initially by a flat battery's being connected to the supply. The fuse also must be able to accept the current taken by the alarm circuit and any ancillary components, such as sounders and strobes, in the event of a power supply failure and the alarm's being activated—a common occurrence when an attacker has switched off the power supply in an attempt to defeat the security system.

Another similar circuit is shown in Figure 8.5a, with the addition of an LED that will warn of supply failure or that the circuit is powered up.

The capacity of the backup battery should be considered, since if it is capable of supplying power to the system for only an hour or two because it also has been powering activated sounders and the like, the system then becomes open to attack if the backup power source is exhausted!

The largest capacity backup battery that can be accommodated inside an enclosure should be chosen, as well as periodic renewal as per the manufacturer's specifications.

APPLIED SECURITY DEVICES AND CIRCUITS

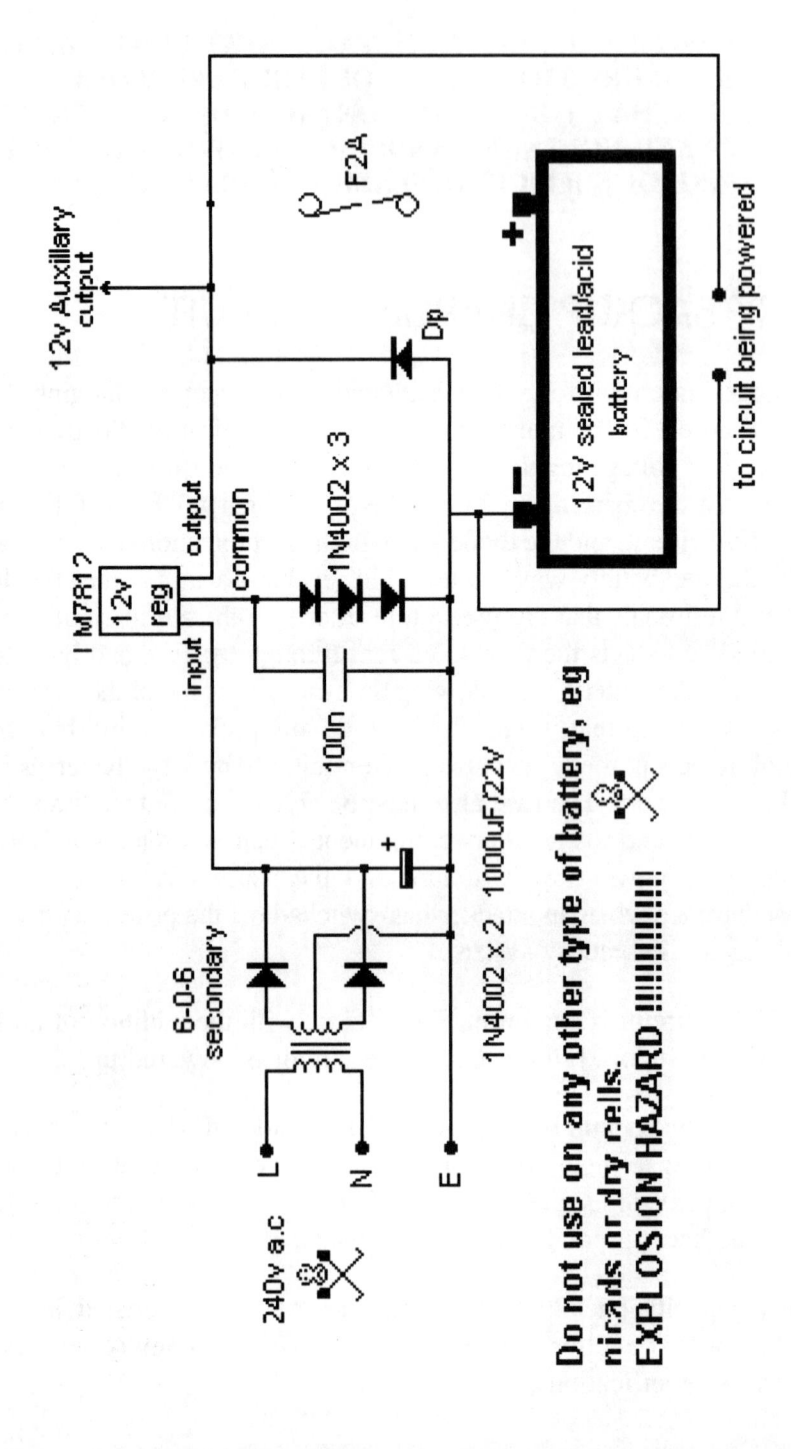

FIGURE 8.5

12 volt lead/acid backup battery charging circuit

Do not use on any other type of battery, eg nicads or dry cells.
EXPLOSION HAZARD !!!!!!!!!!!

MISCELLANEOUS DESIGNS 119

FIGURE 8.5a

Battery-charging circut

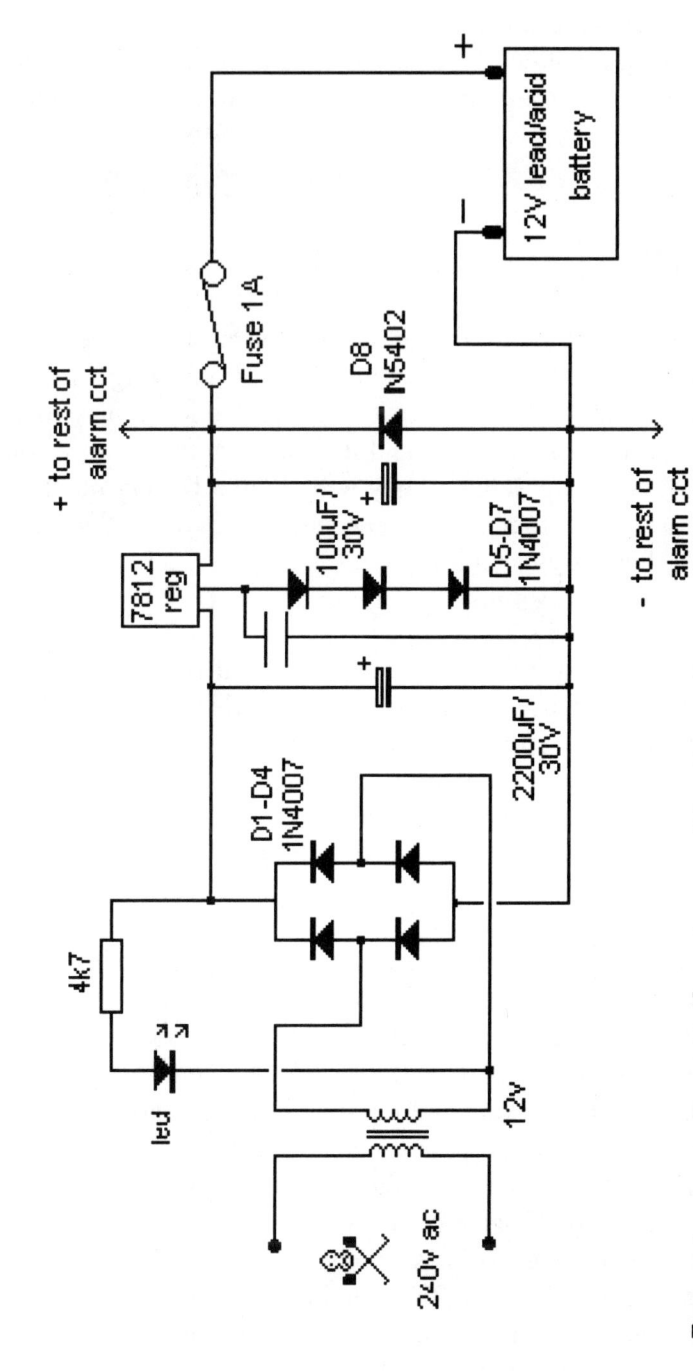

Do not use on any other type of battery, eg nicads or dry cells.
EXPLOSION HAZARD !!!!!!!!!!!

A circuit that allows either a method of systematically or, where possible, continuously monitoring the state of the battery should be employed. A circuit that can do this using a 741 op-amp is shown in Figure 5.1, in Chapter 5.

BRIEFCASE ANTI-THEFT METHOD

Anyone reading this book is aware of the dangers of having property stolen from them and others by the "grab it and run gangs." If carrying a briefcase, you are a target for villains who know the case likely contains an expensive electronic organizer or laptop computer and are unaware that it may contain only lunch and the Sunday supplement; they will attempt to very quickly and forcibly grab the case from your grip. Some people advocate the use of a security wrist chain, but the possible physical damage possibly caused to a victim using such a device does not bear thinking about.

A method of security around for several years is that of placing a very loud alarm device inside the briefcase, so that hopefully the thief will drop the case immediately rather than risk running away past a security person with the alarm sounding. Other devices sometimes are fitted to briefcases—such as highly visible smoke dispensers, high-voltage shocking devices, and the like—although if these devices are triggered accidentally, the results can be alarming to the owner of the device!

A few methods that can be fabricated cheaply and easily are shown in Figure 8.6. The top three methods rely on either maintaining a constant grip on the touch plates or relying on the weight of the case to hold down a pushbutton or tamper switch against the fingers of the carrier. A method I prefer is to utilize a cheap jack plug and socket combination. In this design, the jack plug is connected by a strong piece of cord, or thin chain, to the wrist of the carrier. When the plug is inserted into the jack socket—the sort used for earphones in radios and so on—the internal alarm circuit is armed, preferably with a key switch that also is mounted on the case body. If the jack socket is wired correctly and if the plug is pulled out, then the internal contacts of the socket will make, so triggering an instant alarm. Any of the latching circuits described elsewhere in this book (for example, the one shown in Figure 8.51a,

Briefcase anti-theft Method

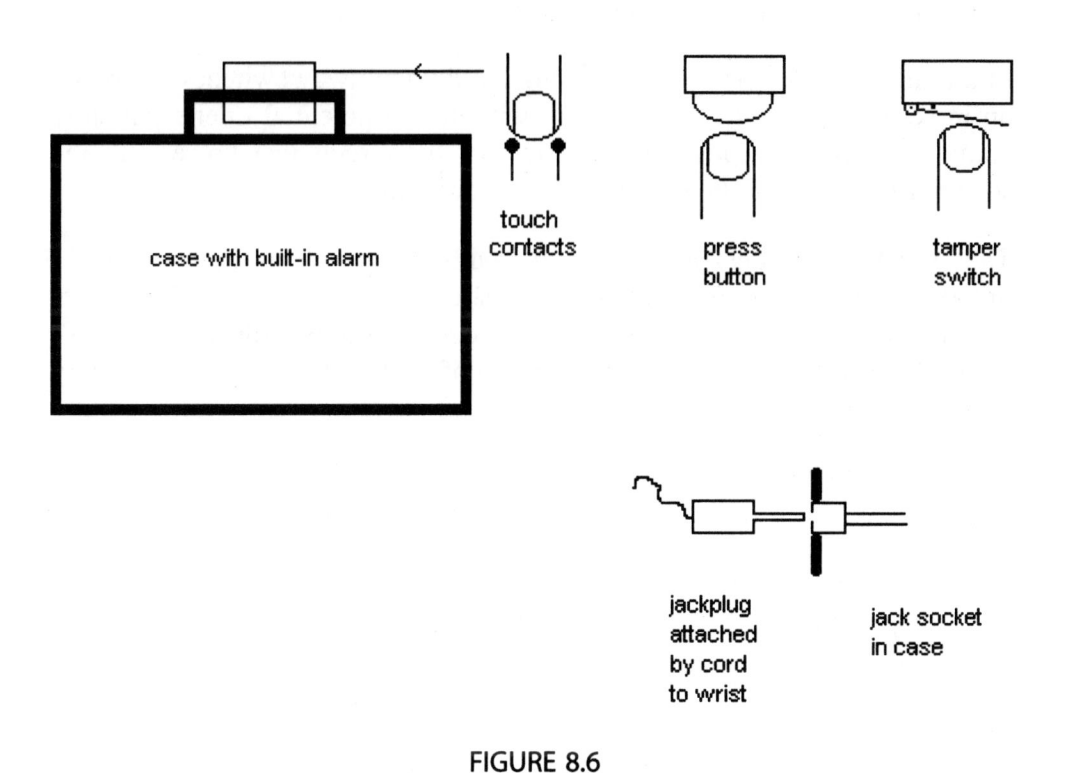

FIGURE 8.6

later in this chapter) may be utilized as part of the design. I recommend that a series of holes be drilled in the case wherever the sounder is mounted so as not to attenuate the noise unduly.

COMBINATION LOCK

There are many situations when a combination lock may be used instead of a conventional lock. The unit described in this section is shown with a supply drawn from a battery; this may be replaced by a ac-powered supply only if a backup battery system is implemented. This is because if there is a power failure, the property will not be accessible, a scenario with dire consequences should there be a fire or similar disaster.

In the most simple form, a row of single-pole, multiway switches are wired up in series (see Figure 8.7) so that, without the potential thief's patiently going through every possible combination, the circuit that powers up the solenoid lock will not be completed in a long time.

The security of this design can be increased by simply increasing the number of switches, including dummy switches, adding a time-delay circuit so that even if a potential thief stumbles across the correct combination, he or she will not realize such. Alternatively, the security of the design can be increased by

"Electronic" Combination Lock

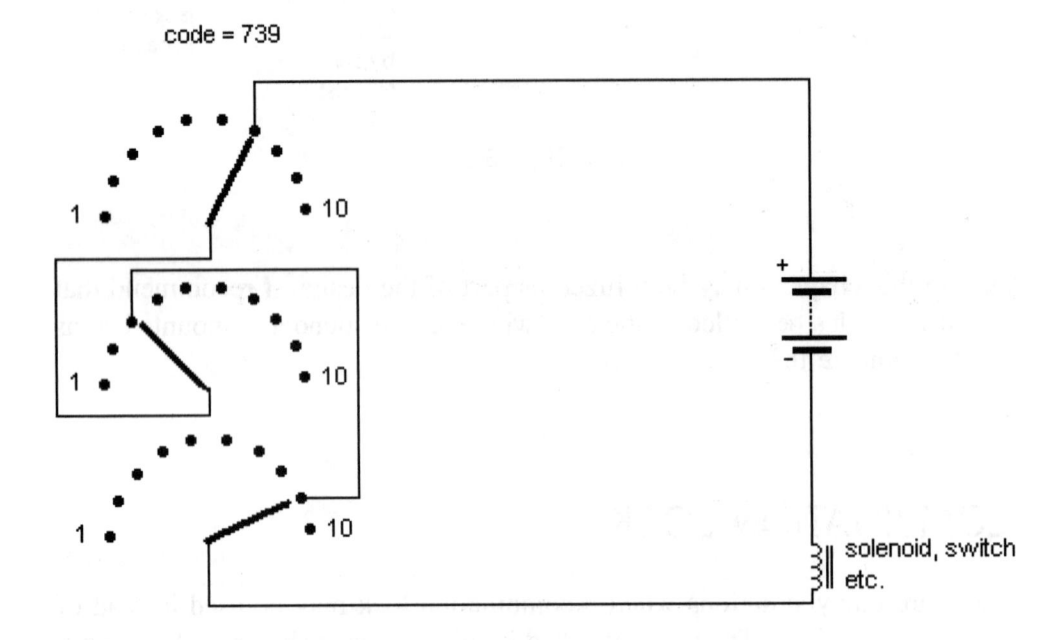

FIGURE 8.7

using the circuit shown in Figure 8.8, which has an additional tryout switch that provides positive-going pulses each time it is pressed. The system alarm is triggered after receiving "X" amount of attempt pulses that are generated by the system, as demonstrated in Figure 8.9. The circuit uses the CMOS 4017 decade counter/divider IC, whose pinouts are shown in Figure 8.10. This IC has 10 outputs, one of which will go high in succession after a positive going pulse is received by the clock input pin. So long as the clock inhibit and reset pins are held low, the IC will count up one with every clock pulse. If, as in the example, four clock pulses, or attempts, are chosen, the four output pin goes high, triggering an alarm. If a latching system is required, the chosen output pin can be connected to the clock inhibit, forcing it high and so stopping the counter from continuing to the next count. The method of fitting a solenoid lock to a door is illustrated in Figure 8.8a.

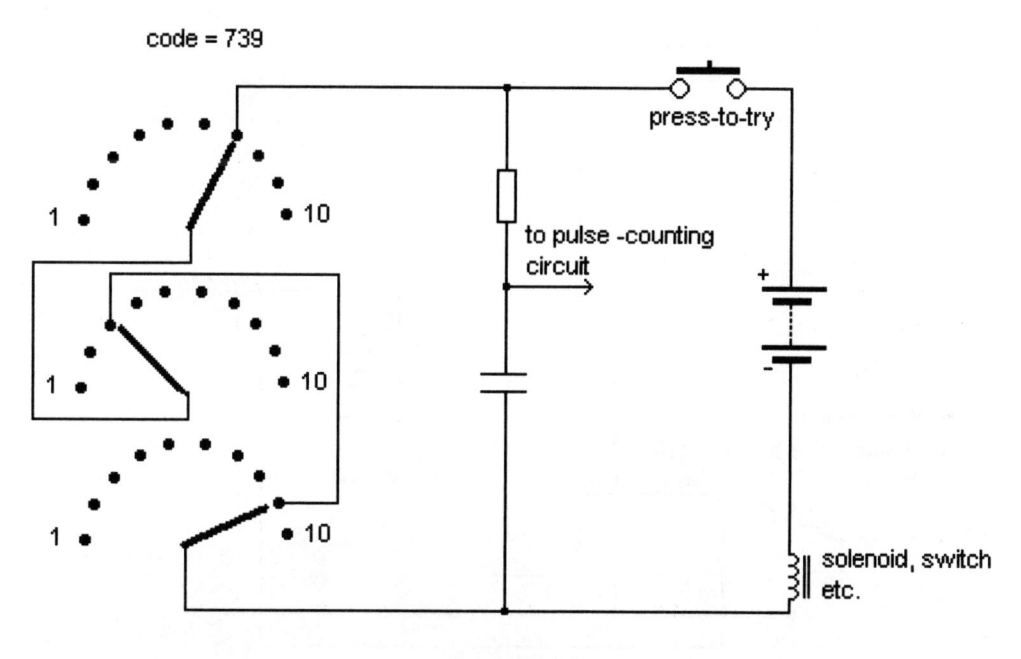

FIGURE 8.8

124 APPLIED SECURITY DEVICES AND CIRCUITS

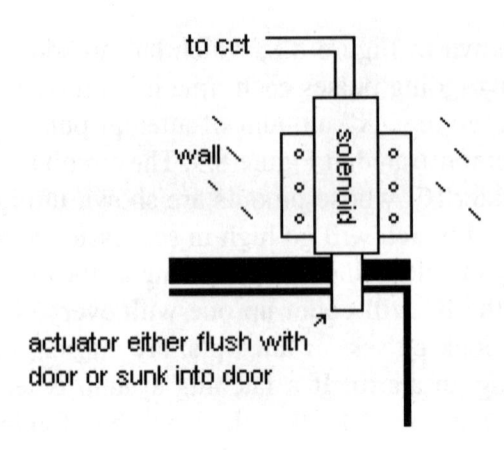

FIGURE 8.8a

CD4017 decade counter/divider ic connected to give output after four pulses

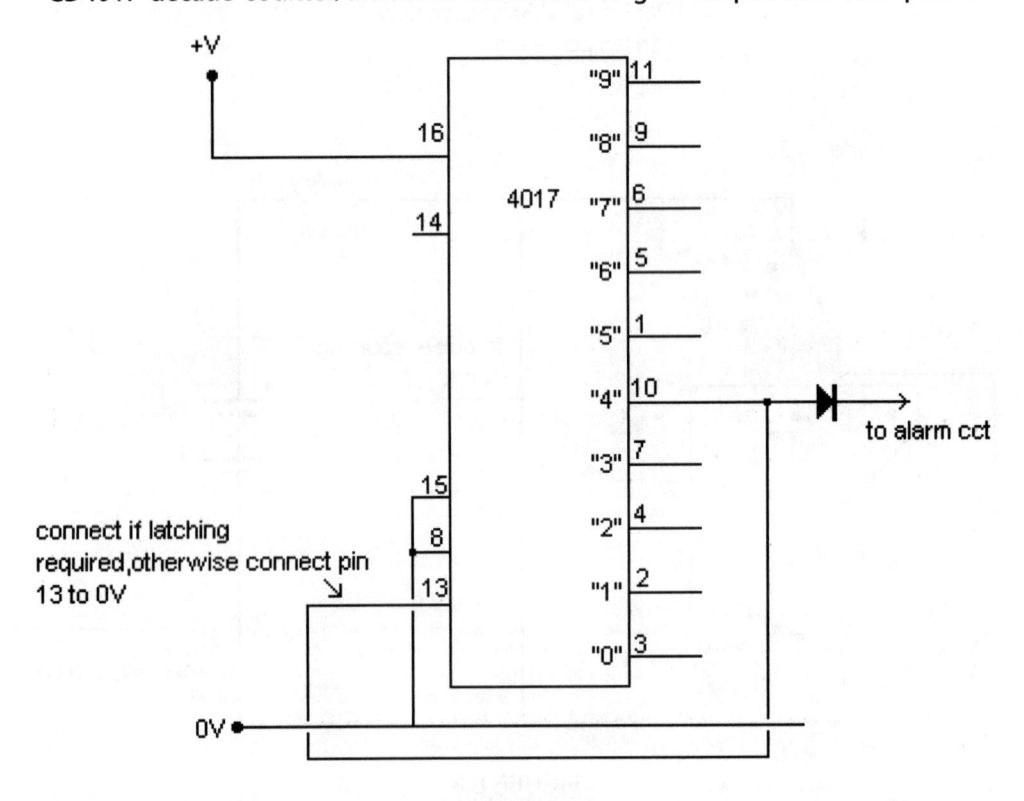

connect if latching required, otherwise connect pin 13 to 0V

FIGURE 8.9

CD4017 decade counter/divider ic

```
              ┌─────────┐
          16  │         │ 11
       +V ────┤         ├──── "9"
              │         │  9
              │   4017  ├──── "8"
          14  │         │  6
     ↑ clock ─┤         ├──── "7"
              │         │  5
              │         ├──── "6"
              │         │  1
              │         ├──── "5"
              │         │ 10
              │         ├──── "4"
              │         │  7
          15  │         ├──── "3"
      reset ──┤         │  4
           8  │         ├──── "2"
         0V ──┤         │  2
          13  │         ├──── "1"
↑ clock inhibit┤         │  3
              │         ├──── "0"
              └─────────┘
```

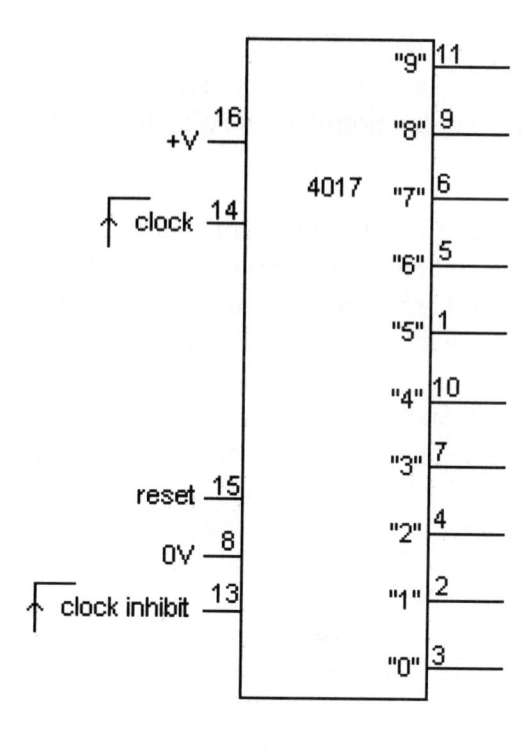

FIGURE 8.10

The position of the actuator is important, ensuring that it cannot be tampered with, wedged in with a piece of cardboard, and so on. If at all possible, with the risk of weakening the door or door frame, it may be wise to sink and hide the solenoid and actuator hole into the frame and door.

CONTROLLING SUPPLY VOLTAGE WITH SENSOR SWITCHES

A method of using normally open or normally closed switches to turn power on to the alarm circuit is not feasible in some circuit designs because of the possible high currents required by the circuit. It is also not practical to have a normally closed switch be open, as in the case of a single-pole magnetic door switch. However, if a simple transistor switch is used, this will mean that the sensor switch is switching only a small amount of current since the transistor is doing all the work; a further aspect of using a switching transistor is that it is possible to use both normally open and normally closed switches in the system.

With reference to Figure 8.11, the transistor, which is effectively in series with the supply voltage and circuit to be powered, does not conduct because there is no voltage to the base of the transistor. If the normally open sensor switch is now attacked—that is, becomes closed—the base will become connected to the supply voltage via the 10k resistor. During the time that the switch is closed, the transistor will continue to remain switched on, so supplying the circuit with power. This means that the circuit will remain powered up only while, say, the door or window is left open, because as soon as the burglar closes the aperture, the alarm will stop; that is, the circuit is not of a self-latching variety.

A method of switching a transistor using a normally closed switching device is shown in Figure 8.11a. In this case, the normally closed switch is grounding the switching transistor base to ground, maintaining a base voltage of 0V. If the switch is attacked, then it will become an open circuit, so the switching voltage

Method of controlling supply voltage with either normally open or normally closed switches

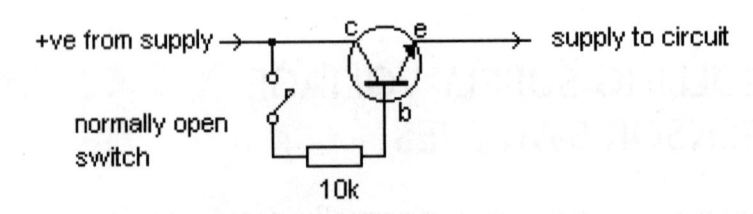

FIGURE 8.11

through the resistor will give a base bias that will turn the transistor on, allowing current to flow through to the alarm circuit being powered. Once again, as in the previous example, if the sensor switch is returned to the original setting, that is, closed by the burglar, then power to the alarm system is shut off.

It is possible to connect more than one of the same type of normally open or normally closed switches together so as to establish a multipoint system, as shown in Figure 8.12 and Figure 8.12a. If a mixture of normally open and

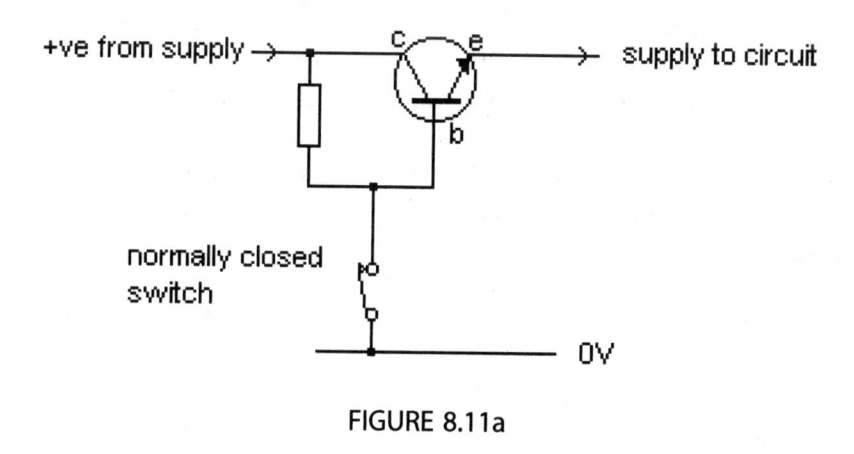

FIGURE 8.11a

Method of controlling supply voltage with either multiple normally open or multiple normally closed switches

multiple normally open switches, connected in parallel

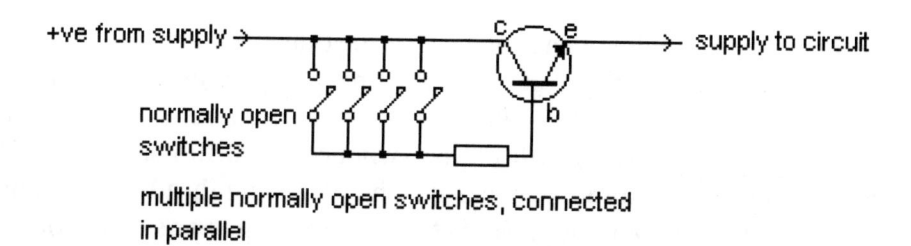

FIGURE 8.12

FIGURE 8.12a

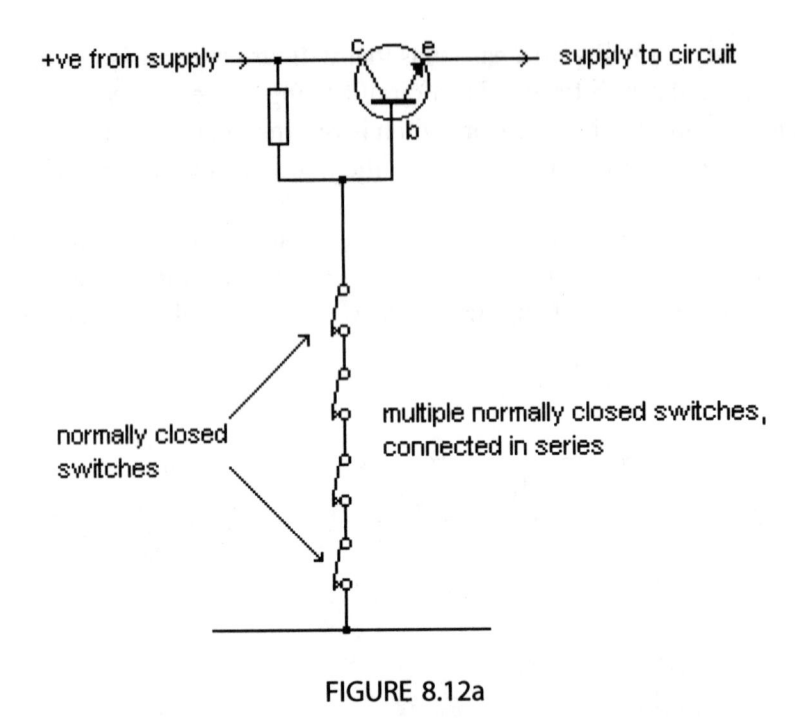

normally closed switches is required, then both circuits would have to be paralleled, as illustrated in Figure 8.12b.

If a switched transistor is used in these configurations, then the current, voltage, and power ratings of the switching device must be considered to prevent any damage to the device, especially since bipolar transistors have a habit of becoming a continuous short-circuit after blowing up, which means that the supply current then would become continuously on.

The design using normally open switches has the drawback that there is a limit on the amount of current that can be drawn through the transistor, because of the current-handling capability of the transistor and because if the base resistor is too large, the transistor will not be switched on to full saturation. The design using normally closed switches has the preceding disadvantages, plus the fact that a small amount of power is taken from the power supply continuously as current leaks through the resistor and through the normally closed switch contact to ground.

Method of controlling supply voltage with both normally open and normally closed switches

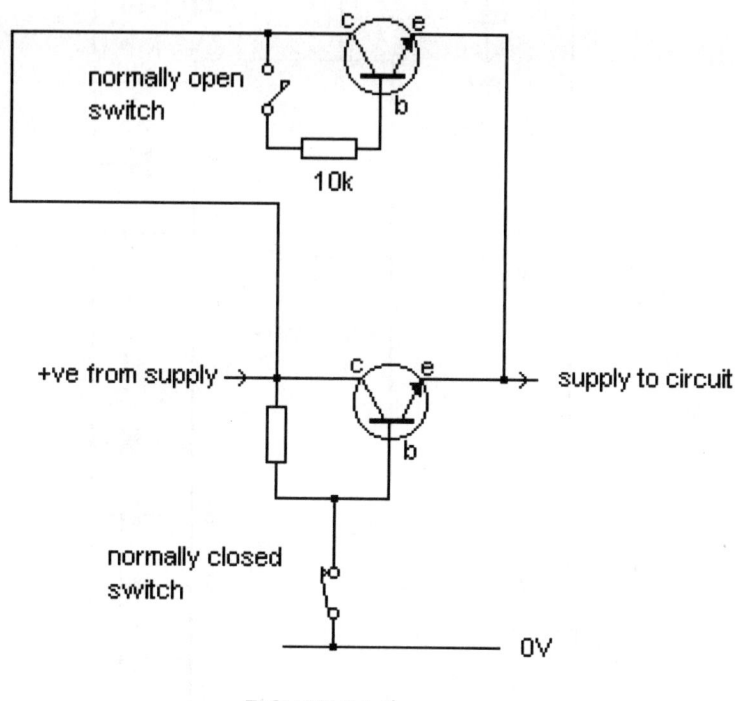

FIGURE 8.12b

COUNTDOWN (OR COUNT-UP) DISPLAYS

An attractive and useful feature of some commercially available alarm systems is an audible beeping that alerts the user of entry or exit delay. The two designs shown in Figure 8.13 and Figure 8.14, however, give a visual countdown to an event by lighting up a row of LEDs to indicate time left before an event takes place.

The first circuit uses an LM 3914 display IC that uses an internal clock whose speed is set externally. The last output in the one shown, pin 10, is connected to the circuit to be armed, maybe a self-latching power-switching circuit.

Countdown using LM3914

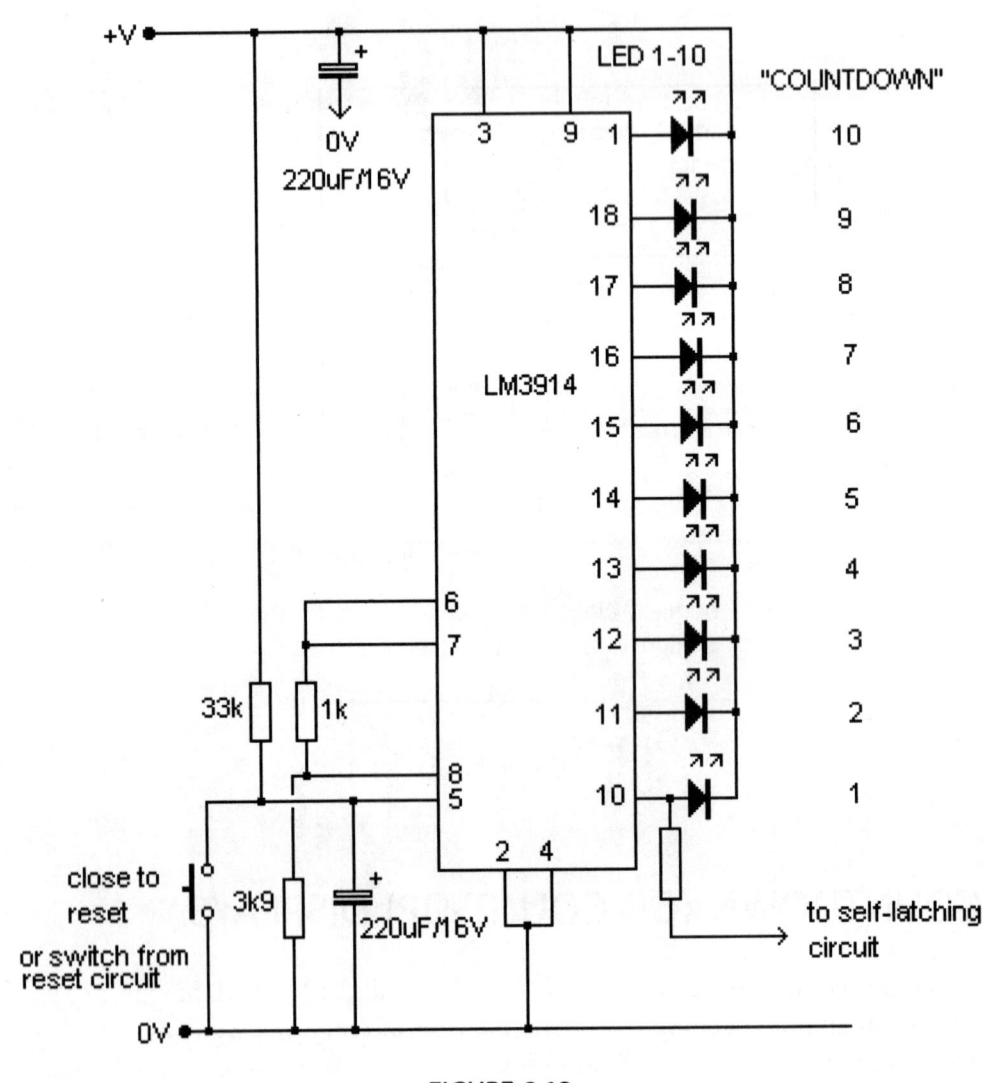

FIGURE 8.13

MISCELLANEOUS DESIGNS | 131

Countdown using 4017

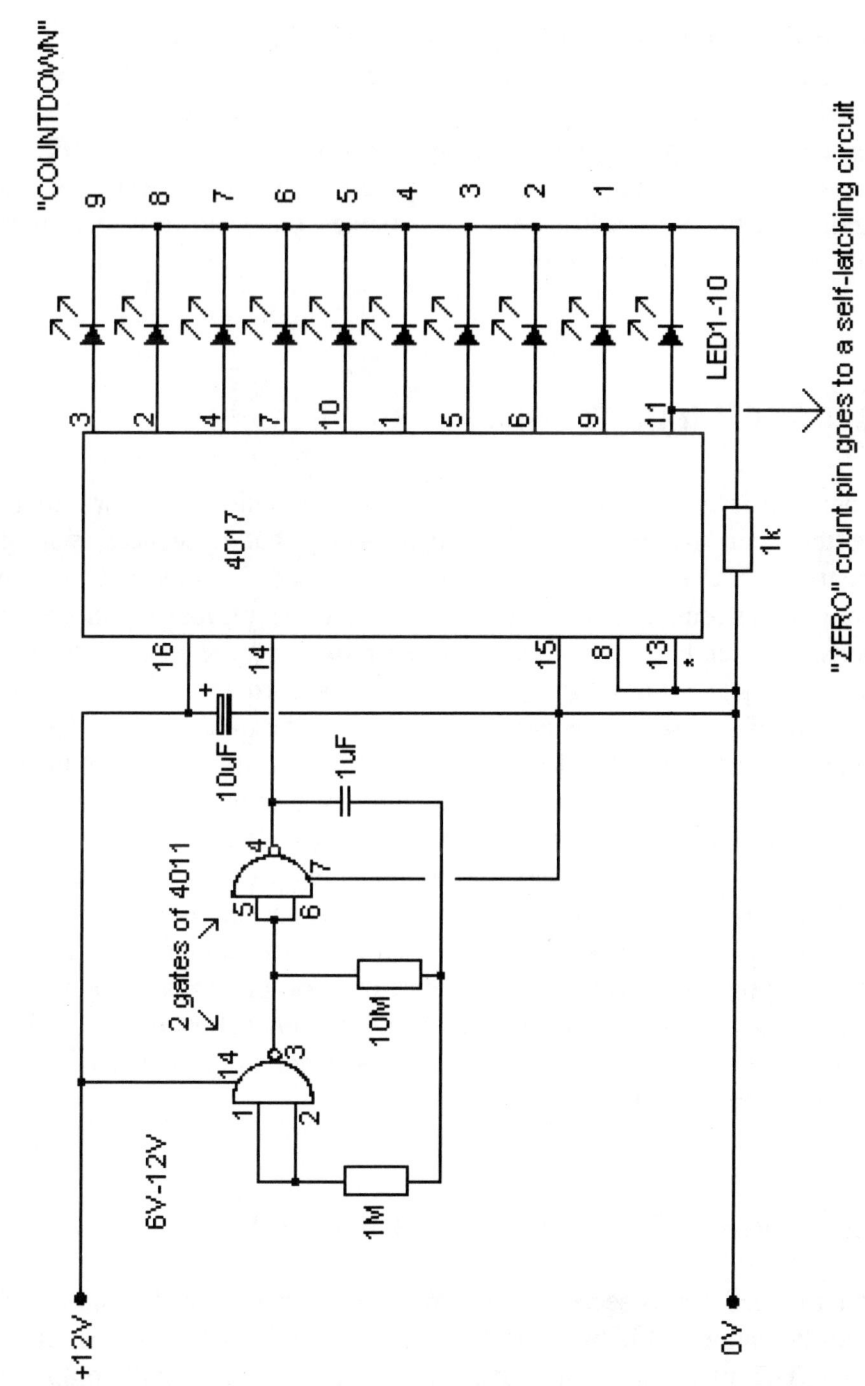

FIGURE 8.14

The second circuit uses a 4017 divide by 10 IC. This IC requires an external clocking circuit, or clock pulse, which in the case illustrated is built up from a pair of NAND gates. The speed of the clock is controlled by a CR combination. On power-up, the clock generates pulses that trigger the 4017. Each of the 4017 outputs goes high one at a time only until the desired count is reached, with the chosen last output pin being connected to a self-latching circuit, as in the previous example.

DISCRETE MONITORING

It sometimes is required to use some kind of method to monitor situations where the interested party requires information that it suspects will give information of a crime being planned or discussed. To this end, it then is required that information be obtained by using a surreptitious system, such as methods of listening to telephone conversations and the like. Note that in many localities and states, the use of devices to perform this type of operation, or being in possession of any equipment that may be used for this method, is illegal, with severe penalties for anyone caught doing so. Anyone considering constructing or using such equipment should first thoroughly familiarize himself or herself extremely with local and national legislation. Remember that your telephone company will have far more sophisticated fault-finding equipment than you may care to imagine. A faulty line caused by your clandestine installation will not be treated lightly; even a standard maintenance-line sweep could bring your modification to the notice of important organizations. In the United Kingdom, for example, it is an offense to connect unauthorized equipment to the phone system, which often falls under the category of stealing electricity. You have been warned!

Telephone–to–Tape Recorder Interface

The first device in this section is shown in Figure 8.15, which is an interface that can be connected between a telephone line and a standard cassette tape recorder. The microphone input of a tape recorder is very small compared to

Telephone to tape recorder interface

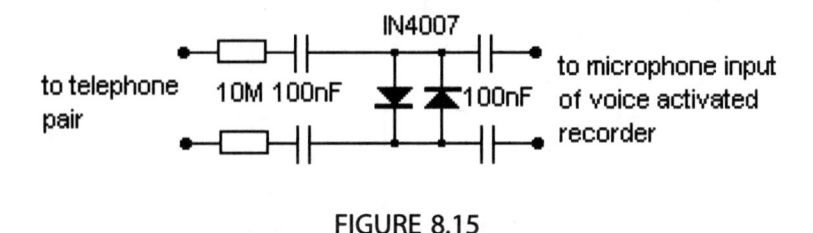

FIGURE 8.15

the amount of audio level and voltages that appear on a telephone line, which swings from around 10V when the phone is off the hook to 50V when on the hook, to a voltage of around 150V or more peak to peak during ringing. To provide adequate protection to the recorder circuitry, the signal must be dramatically attenuated by large-value resistors, DC blocked by capacitors, and, to add an extra amount of protection, a pair of back-to-back diodes to prevent the voltage's exceeding 0.7V. The circuit is intended to connect to a VOX (voice-operated switch) style of recorder, but for a normal style of recorder, the 10M resistors may be decreased in value to around 100k. To obtain correct audio level, you will need to adjust these values to suit the particular recorder being implemented; with too low a value, the recordings will be distorted, and with too high a value, recordings will not be loud enough. If a VOX recorder is used, the recorder switches on and off automatically whenever speech is picked up, so as to conserve tape space.

Telephone Line In Use Indicator

Figure 8.16 shows an indicator that is connected to a telephone wire pair so that an LED is illuminated whenever a telephone handset is taken off the hook. This design is useful if someone on the same line is surfing the Internet, where if another handset on the same line is picked up, a long and expensive upload or download is corrupted. The circuit is polarity-sensitive, so you must

Telephone line in use indicator

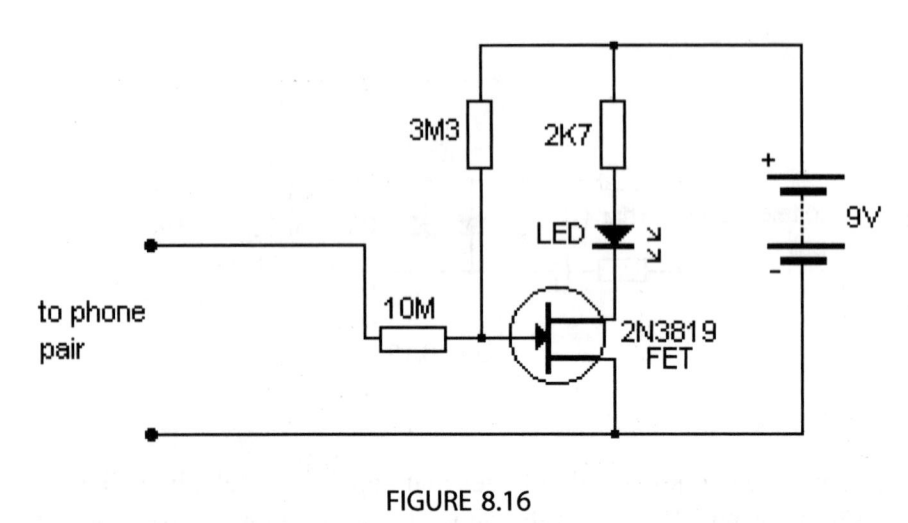

FIGURE 8.16

use trial and error to find out which way round the two phone-line wires must be connected. The unit then can be finished off with a smart case, battery switch, and a flying lead terminated with a telephone plug.

Telephone Line Tap Detection

A detector circuit that requires no separate power supply—that is, a parasitic device obtaining power from the telephone line—is shown in Figure 8.17. The circuit will give an indication if a second extension telephone handset is lifted during a phone conversation, as well as give an indication if a crude, voltage-dropping tap has been placed on the line (but only after the detector has been installed!). To install and set up the device, first make a break in one of the phone-line pairs. You can find the pair by finding which two wires have around 50V DC across them, falling to around 10V whenever a handset is picked up. Once connected, the telephone handset is lifted, and the pot is adjusted so that the green LED just comes on. If another handset on the same line is now lifted, the red LED should come on. You can

Telephone Line Tap Detection

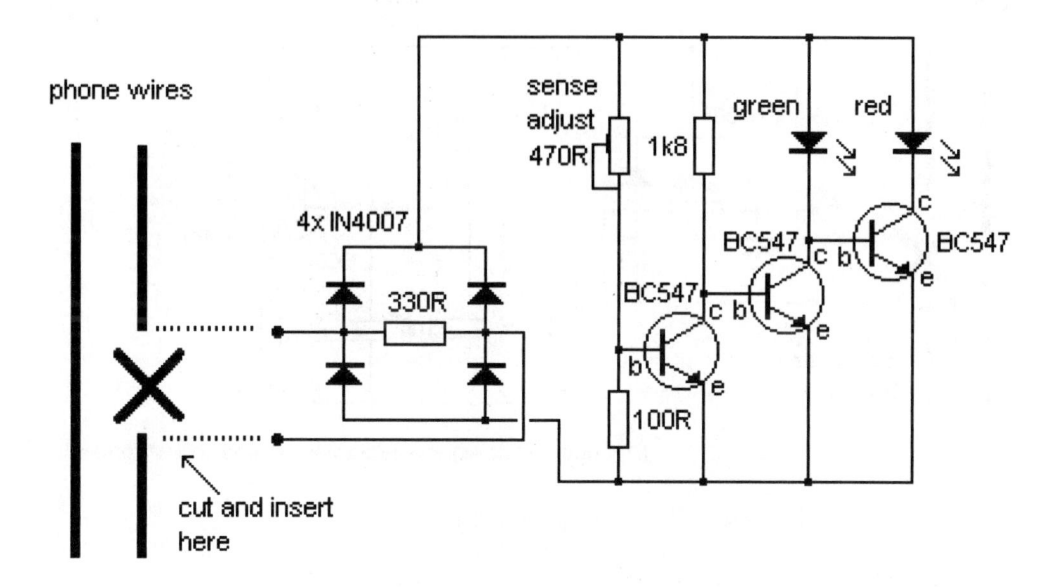

FIGURE 8.17

improve the sensitivity of the circuit settings by replacing the pot with a multiturn device.

Telephone Transmitting Device

A circuit that is parasitic, with a range of a few hundred meters, is shown in Figure 8.18. The device can be built into a very small space or hidden in a telephone base, telephone pole, wall socket, and so on. The unit will transmit on around 100MHz and operate only when any telephone handset with which it is in series is picked up. Tuning is performed by adjusting the coil, by stretching or compressing it, or by replacing the tuning capacitor across the coil, L, with a small-value trimmer capacitor. Because the RF energy usually finds a way into the telephone system wiring, even with the RF decoupling

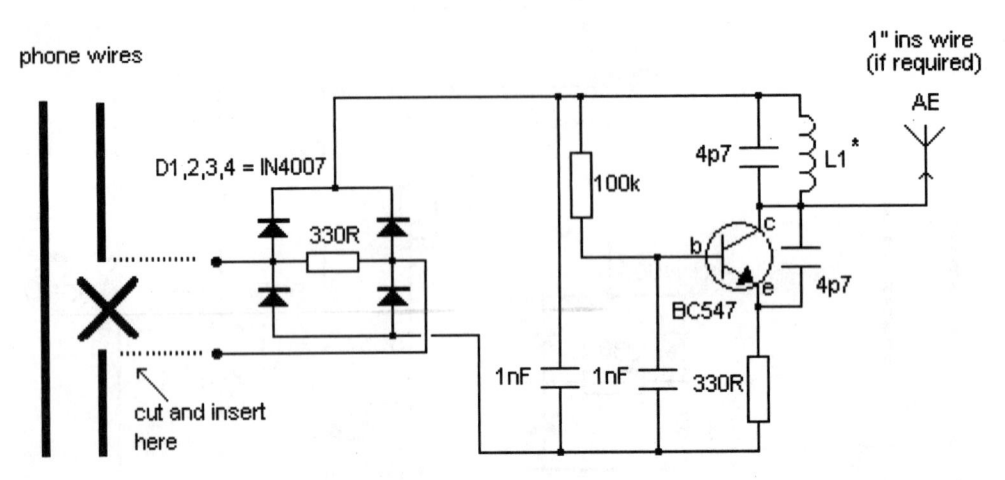

FM VHF Telephone Transmitter

FIGURE 8.18

L1 = 6 turns 22swg enamelled wire wound on 1/4" former

capacitors that are connected as shown, no aerial length is required. Output is received on a standard VHF-FM–type receiver.

Voice Transmitting Device

An almost identical free-running VFO circuit as the previous design, but with a microphone attached, is shown in Figure 8.19. Current consumption is in the order of around 20mA, and with an aerial constructed from around two feet of ordinary insulated wire, a range of a quarter of a mile without obstructions may be possible. As with RF designs, wiring should be kept as short as possible, especially battery leads since feedback will cause the device to malfunction.

Automatic Recording from Receiver

This device is plugged into the earphone socket of a radio receiver, with outputs that go to a standard cassette recorder so that the tape recorder will

Micro Voice Transmitter with minimum parts count

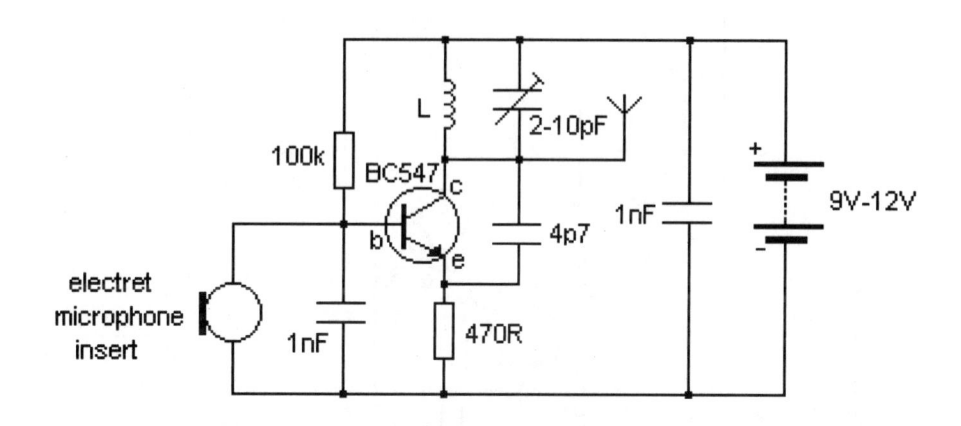

FIGURE 8.19

L=6 turns 22 gauge enamelled wire wound on 3/16" former

switch on and record only if the receiver picks up a signal, say, from one of the transmitters mentioned previously. The circuit is shown in Figure 8.20. The idea of this device is that the security operative can hide the receiver/recorder nearby to the transmitter and then go back at some later date to replay any conversations that may have been transmitted. It is crucial that the output frequency of the transmitting device be selected carefully, on a quiet frequency allocation, so that the receiver does not hear a commercial radio station's output; otherwise, the tape will not last long!

Automatic Recording from Telephone

To save tape space, and in the absence of a VOX-type recorder, the design shown in Figure 8.21 can be used. The PNP transistor is biased when the phone is on the hook so that the output transistor remains off; therefore, the

138 APPLIED SECURITY DEVICES AND CIRCUITS

Automatic radio to cassette recorder

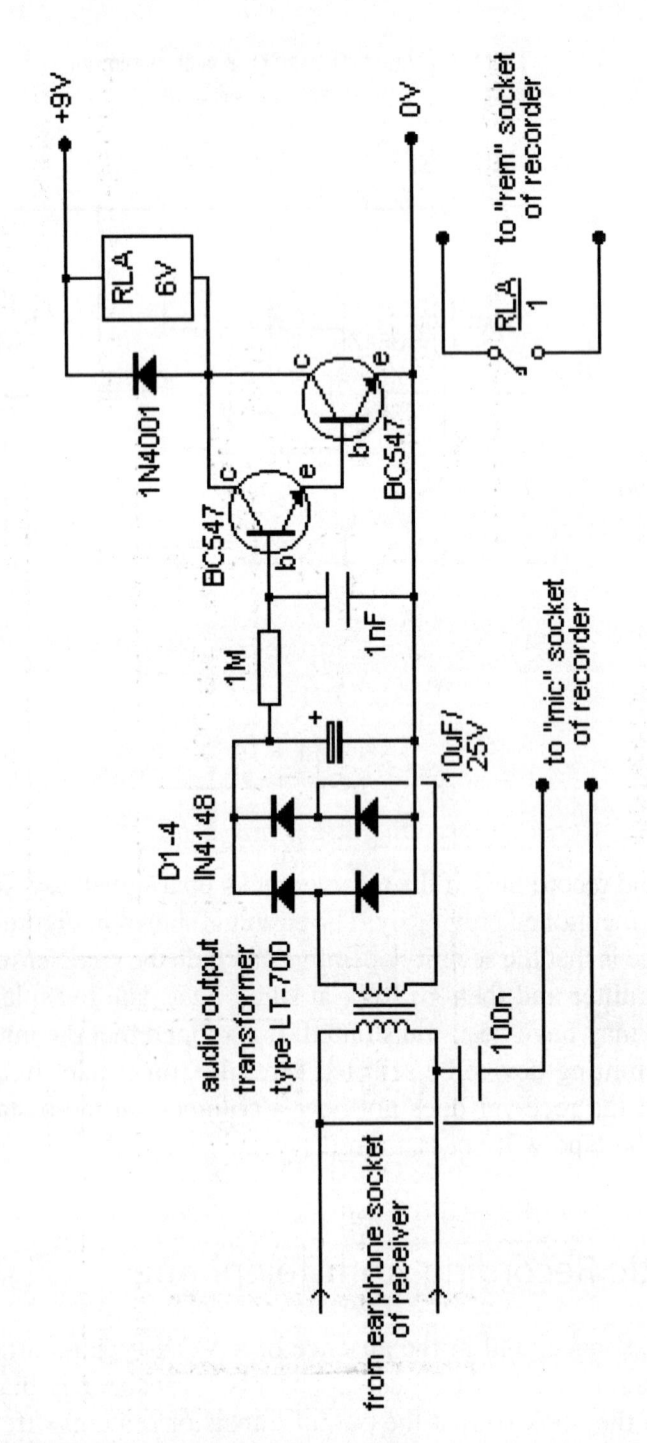

FIGURE 8.20

MISCELLANEOUS DESIGNS 139

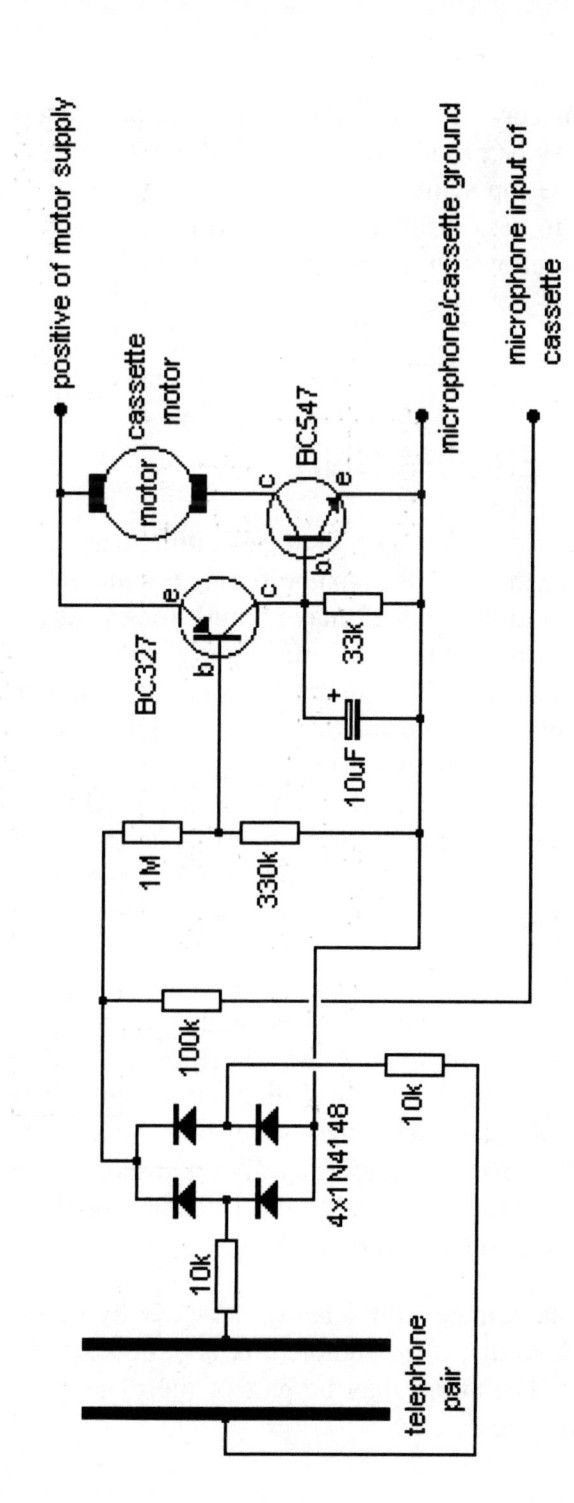

FIGURE 8.21

Auto on/off telephone to recorder interface unit

cassette motor stays off until the handset is picked up, switching on the recorder. It may be prudent to insert a DC blocking capacitor in series with the microphone input, of a value of around 100nF, which still will allow audio signals to pass through. The electrolytic capacitor introduces a short hang-time that allows the switch to stay on during short pauses in between speech.

DOOR INTERCOM UNIT

If it is not possible to view a caller at a dwelling to ascertain his or her identity prior to opening the door, the next best thing is some form of intercom device. The unit shown in Figure 8.22 may be built for use as a door intercommunication device or as a house-to-workshop system so that the family members may communicate quickly in an emergency. The system has two components: the slave or remote section, and the master section, which is situated inside the dwelling. Connection between the two units is by two-core screened wire, although ordinary three-core flex may be used if RFI (radio-frequency interference) does not cause a major problem. A spring-biased switch should be used in the master unit so that the major embarrassment of leaving the master unit continuously transmitting internal noises to any passerby on the street does not accidentally occur!

When the push-to-call button is operated, the remote unit internal battery powers the master unit, even if the on/off switch is open. With the ganged switch in the position shown in the diagram, the three-transistor audio amplifier drives the speaker inside the master unit. When the ganged master unit switch is flicked over, the master speaker is disconnected from the amplifier output and is connected instead to the amplifier input, with the reverse happening to the speaker in the remote unit.

To prevent the remote unit's being attacked by vandals, you can add a tamper switch so that if the remote box is pulled off the wall or opened, an alarm sounds. The alarm may be part of the existing dwelling alarm, or a stand-alone device.

MISCELLANEOUS DESIGNS

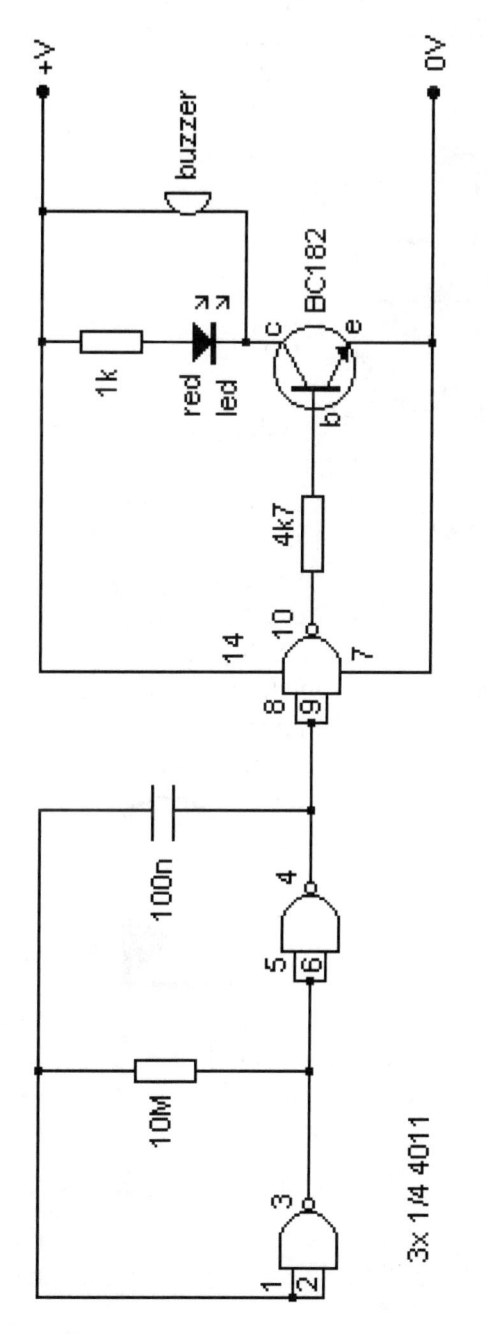

FIGURE 8.22

One second Alternating Flasher / Buzzer using the 4011 ic

3 x 1/4 4011

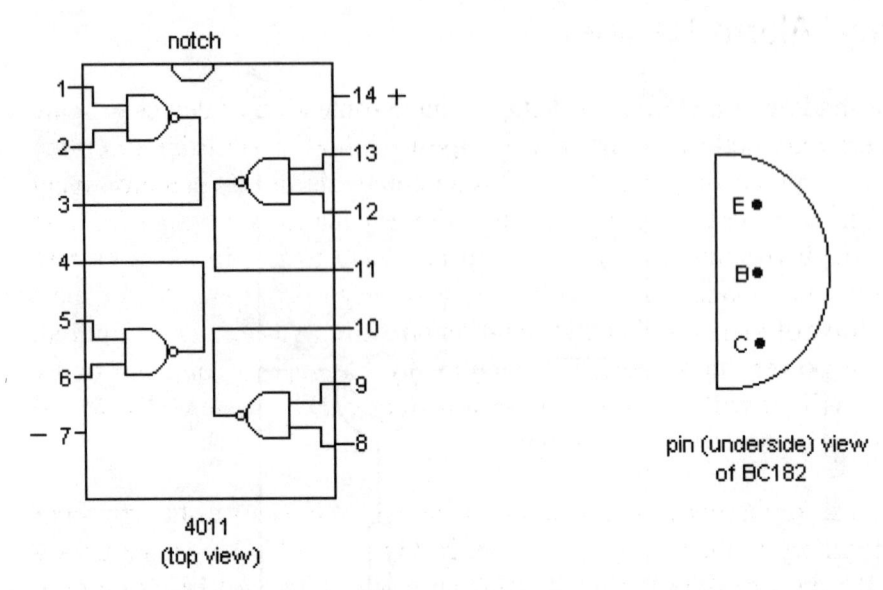

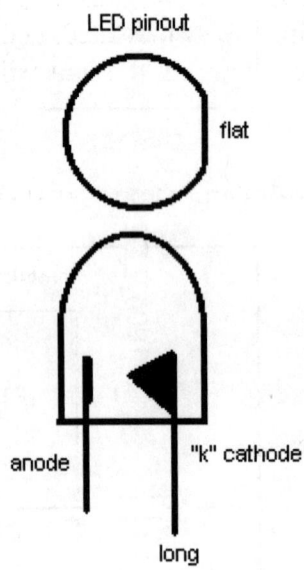

FIGURE 8.22 (Continued)

"Dummy" Alarm Devices

Without a shadow of a doubt, it is better to have some kind of device to scare off intruders than nothing at all. The cheapest method is to install a dummy device, which, in the simplest of forms, is an empty alarm box in a prominent position. The next best thing is to install a dummy alarm box with a flashing light. Although flashing LEDs are cheap and easy to obtain, their current consumption may make them unsuitable for battery operation, which means that some form of mains wiring and step-down/rectification circuit is required. Also, if the power supply is cut off, either by the supply company or by intruders, the light will stop flashing, so a battery backup system also would have to be incorporated in the design.

With all these problems in mind, it would be wise either to install a proper system or simply build the device shown in Figure 8.23. The device uses a common IC, the LM3909 flasher IC (this chip will drive an LED from only 1.5V!). The circuit is wired up as indicated, and if two or more of the largest top-quality 1.5V batteries are used, the unit will happily flash away for two

Dummy alarm box flasher using LM3909

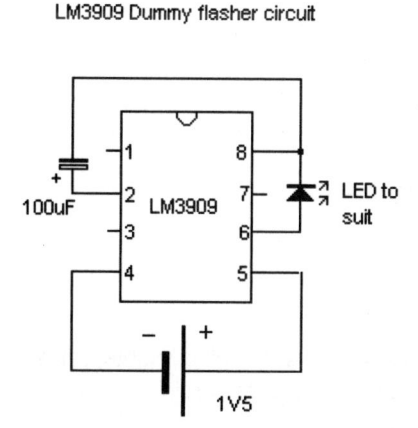

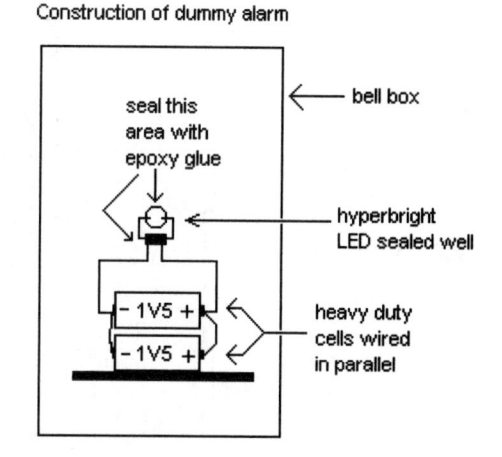

FIGURE 8.23

or three years without maintenance. Corrosion of the batteries and connecting leads is the enemy of this device, so wherever possible, liberally use some kind of sealant or potting compound. The LED must be large, with a wide-angle view and of the hyperbright variety, so it can be seen even during low-level daylight. For anyone who does not want to wire up a proper system, cannot afford it, or is in temporary housing, this device is ideal, because all that is required to remove it and take it to the next lodging is a screwdriver!

The same circuit can be installed inside a small box with an on-off switch and powered by the small penlight or button-cell type of battery, so that the unit can be left on a car dash or placed on the window sill of a caravan, house, etc., at nighttime, to deter burglars or attackers.

A different circuit is shown in Figure 8.24. This circuit may be used if the LM3909 is not available. The circuit requires a supply voltage of 9 to 12 volts, but current consumption is not as low as for the previous circuit. This circuit does have the advantage that a standard LED will flash much brighter than normal, because the capacitor dumps current through the LED via the 10R

Low power consumption flasher using SCR

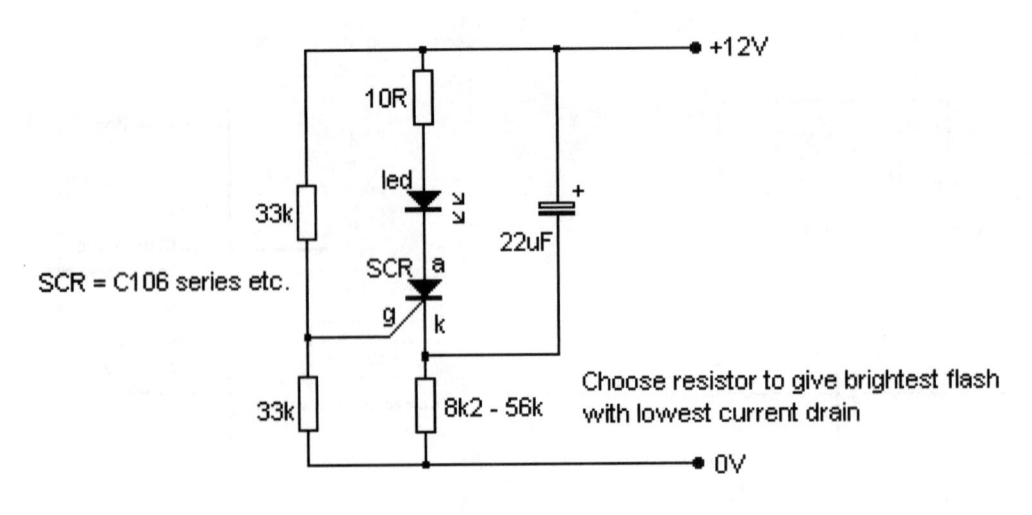

FIGURE 8.24

resistor for a very short on period. Some experimentation will be required with component values in order to obtain the required flash rate and intensity of flash against current consumption.

For a different effect, the circuit shown in Figure 5.8 in Chapter 5 can be incorporated into a dummy alarm box design.

If ac-powered operation is required, see Figure 8.24, a simple power supply to low-voltage circuit.

Take care that there are no possible hazards of fire or electric shock. All metal parts should be well grounded, and high-voltage components should be indoors, well away from the weather.

A further flasher circuit, this time utilizing two transistors configured as an oscillator, is shown in Figure 8.25. Current consumption drawn from the PSU (power supply unit) can be very low, with component sizes not that critical. You can minimize power consumption by increasing the resistor in series with the LED, at the cost of reduced brightness. As with any LED circuit, it may be wise to pay just that little bit extra to purchase one of the many high-output

Simple low-current Mains PSU

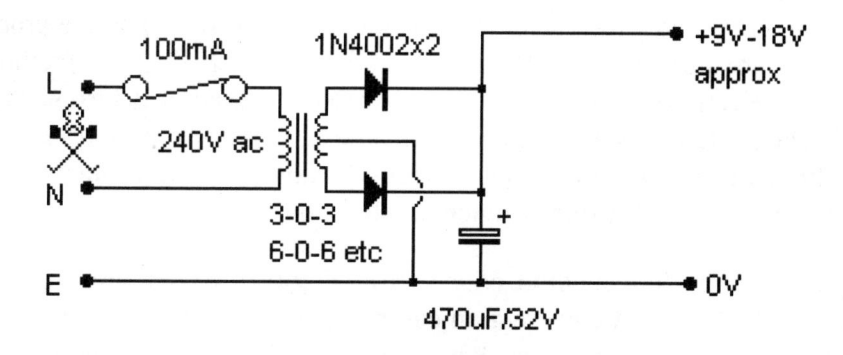

FIGURE 8.24a

Low current flashing led alert

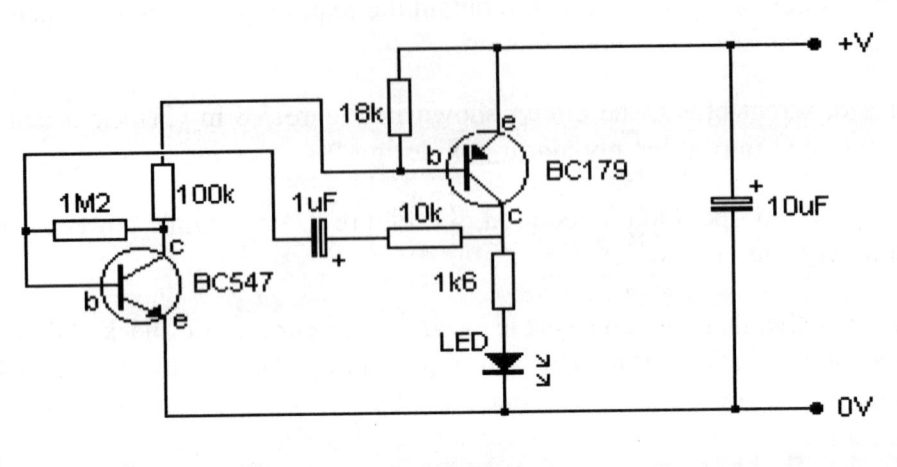

FIGURE 8.25

devices available on the market, since the constructor is looking for maximum visibility during sunlight as well as at nighttime.

ELECTROMECHANICAL HIT SENSORS

A method of detecting whether a vehicle that is backing into a very tight garage space is about to hit the rear wall is often required, especially with cars becoming larger and garages getting smaller. You can overcome the problem by having an ultrasonic system installed, but a much cheaper method is building an electromechanical sensor similar to those used in electronic "maze mice." The switch sensor described here relies on a spring, connected to the outer shielding conductor of a small length of cable, making contact with the inner conductor of the cable if touched, albeit only slightly.

A typical construction method is shown in Figure 8.26, Figure 8.26a, and Figure 8.26b. A rugged construction can be made from shielded RF connectors, such as F-type plugs and sockets used in satellite receivers, although the more slightly fiddly BNC types may be used. A piece of shielded cable, with

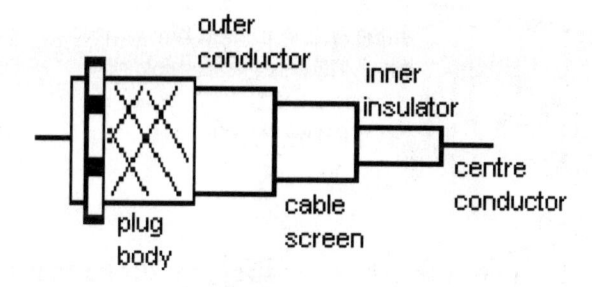

Construction of a home-made Electro-mechanical Hit Sensor

outer conductor
inner insulator
centre conductor
cable screen
plug body

FIGURE 8.25

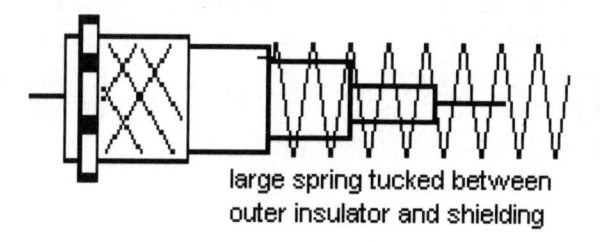

large spring tucked between outer insulator and shielding

FIGURE 8.25

a solid center conductor, is connected to the plug and then stripped back, as shown in Figure 8.26. A length of spring is then pushed over the cable, ensuring that one end of the spring is secure and electrically making contact with the outer shielding. You can fabricate suitable springs from a retractable ballpoint pen spring, or from a cheap, refillable cigarette lighter with a spring that holds in the replaceable flint. You can increase the sensitivity of the sensor by proportionally increasing the length of the spring, which incidentally should have some kind of anti-scratch block as shown in Figure 8.26b, to prevent damaging the vehicle's paintwork.

The RF plug is screwed into a socket that is mounted securely on the rear wall of the garage, where a simple indicator circuit is triggered by touching the

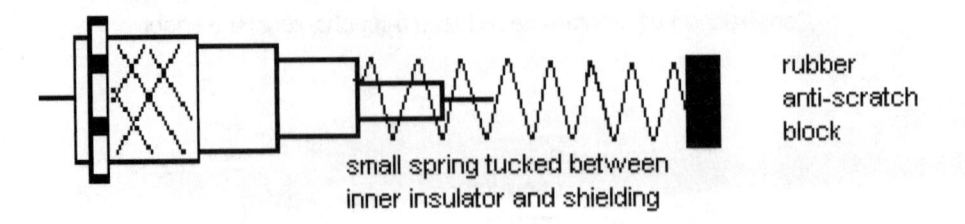

FIGURE 8.26b

spring. The circuit may use a buzzer, siren, or strobe that is situated in full range of the driver.

If the vehicle is of a commercial type, it may be allowed by law to connect the sensor to the actual vehicle, with an output indicator in the cab. If a problem occurs with blind spots—areas not covered by a few sensors—you may use another method, where tamper switches of the type fitted with little rollers at the end of the levers are used in conjunction with free-pivoting wire sensor arms. See this method in Figure 8.27.

EMERGENCY POWER SUPPLY FAILURE BACKUP LIGHTING SYSTEM

Every reader has experienced a mains supply "blackout" at some time or other and knows of the distress, panic, and dangers that can be caused by the same. Many people, especially the very young and very old, find it acutely disturbing to be plunged suddenly into darkness. Although most people are aware of the safe practice of not moving from the safe position they were in when the lights went out, after a minute or two, the dangerous search for a flashlight, probably containing worn-out cells, or a candle, last seen on top of a birthday cake, plus some way of lighting the candle, begins. Furniture and stairs become major stumbling blocks, literally, if a person tries to navigate his or her home in total darkness. Some form of emergency lighting is advisable in every dwelling; although most commercial buildings by law have illuminated

Construction of a home-made Electro-mechanical Hit
Sensor using a rollered microswitch

expected direction of force

anti-scratch block

wire, eg coat hanger

roller

tamper switch

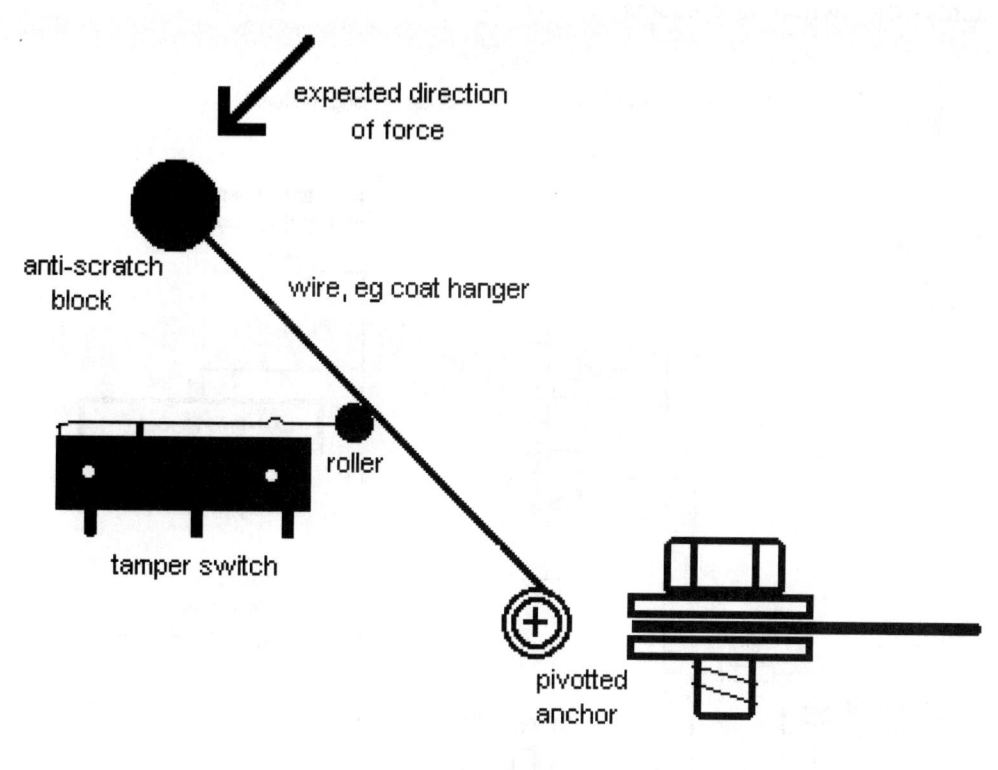

pivotted anchor

FIGURE 8.27

emergency exits, they too would benefit from some level of illumination, especially if a fire was spreading through the building.

A typical emergency power failure backup lighting system is shown in Figure 8.28. The system is based upon the backup battery charger design shown earlier in this chapter in Figure 8.5, but with extra circuitry. The rectified and smoothed supply constantly feeds a transistor-switched relay, and a green LED shows that the power supply is switched on. A commercially available lighting module, which is a 12V fluorescent or incandescent unit, is connected via a set of relay contacts (normally pulled open while ac voltage is present)

APPLIED SECURITY DEVICES AND CIRCUITS

to the battery, which is continuously on charge until the power supply fails. Some way of periodically testing both the battery and emergency low-voltage lighting module is required, so a push-to-test switch is connected in parallel across the switching contacts.

Emergency mains failure back-up Lighting System

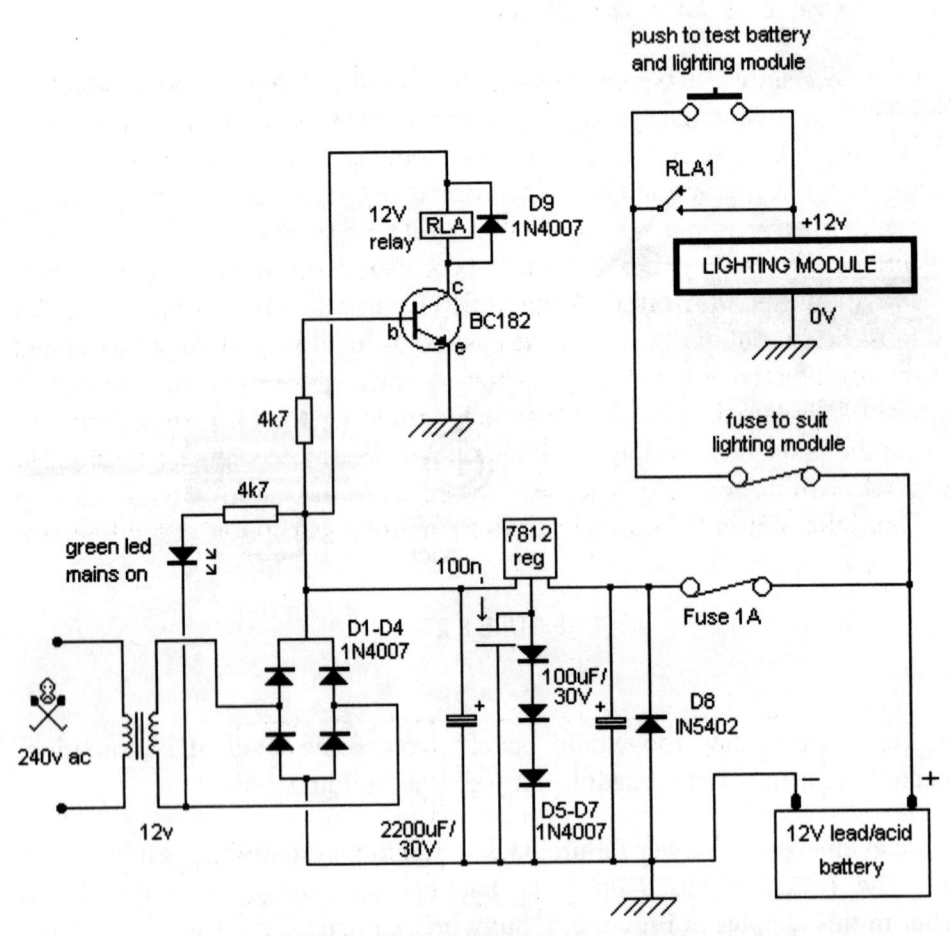

Do not use on any other type of battery, eg nicads or dry cells.
EXPLOSION HAZARD !!!!!!!!!!!!!!

FIGURE 8.28

FISHBITE ALARM

An alarm telling a angler that a fish has been hooked is a positive boon when catches are few and far between, or at nighttime, when a standard float is not easy to see. The device described in Figure 8.29 is a standard two-transistor multivibrator circuit with an audio frequency output from a small speaker. You can alter the audio frequency output by changing the value of the capacitors or the value of the base resistors.

Figure 8.29a shows a typical construction of the fishbite alarm, which is enclosed in a small plastic case. If you use a tamper switch with a long arm, the arm must be covered in heat-shrink sleeving to prevent chafing of the fishing line. If a long-armed tamper switch is not available, you can fabricate an equally good sensor arm by using thin steel wire such as an old coat hanger. This line arm is fixed at one end by means of a small nut and bolt with washers so as to allow free movement of the arm. The arm then passes through a slot that must be cut out of the side of the case so as to allow sufficient movement of the arm to activate the tamper switch. A small groove cut into the end of the arm of the tamper switch will act as a guide for the line arm in order to prevent the latter from sliding to either side of the arm. A bank stick threaded receptacle will allow the unit to be screwed on to the end of a bank stick so as to hold the unit in the correct position. With the exception of the line arm

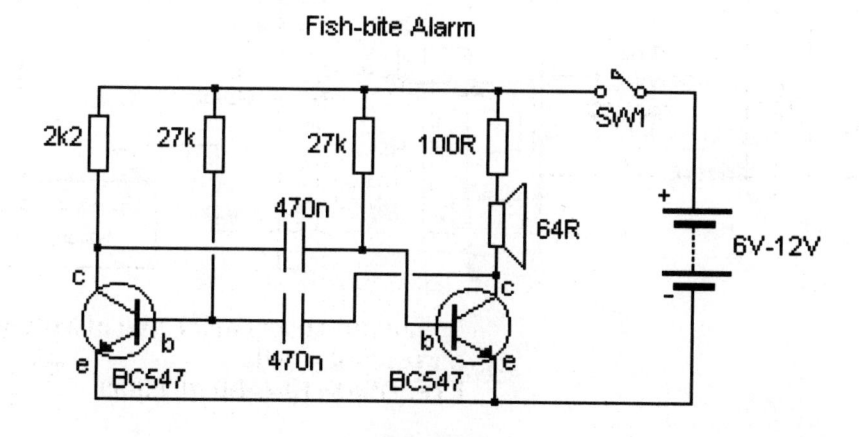

FIGURE 8.29

slot in the side of the unit, the device is relatively shower-proof, although you may want to glue a shield over the slot so as to act like an umbrella, with a drain hole or two drilled in the bottom of the case.

After being threatened by a (200-pound) poor loser–fellow angler during a fishing match about the noise generated by the bite indicator, I decided to add a visual indicator to the alarm. Figure 8.29b shows a quick modification to the circuit that lets the angler switch the audible alarm over to a visual alarm by using a two-way toggle or slider switch that will disconnect the speaker and series resistor and connect an LED and current-limiting resistor.

FLOODLIGHT SWITCHING ALARM INDICATOR

It is all well and good having an alarm system installed, but if some method can be used for attracting attention other than a standard bell and strobe—or, as in many cases, just a bell/sounder—all the better. The device shown in Figure 8.30 is only a suggestion the designer can modify to suit requirements.

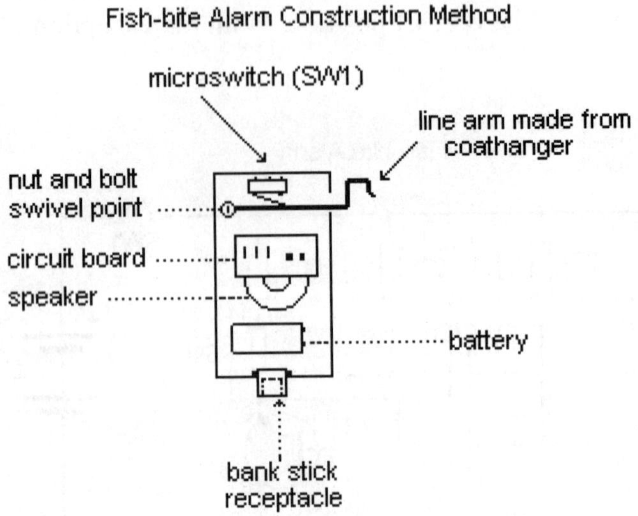

Fish-bite Alarm Construction Method

FIGURE 8.29a

MISCELLANEOUS DESIGNS 153

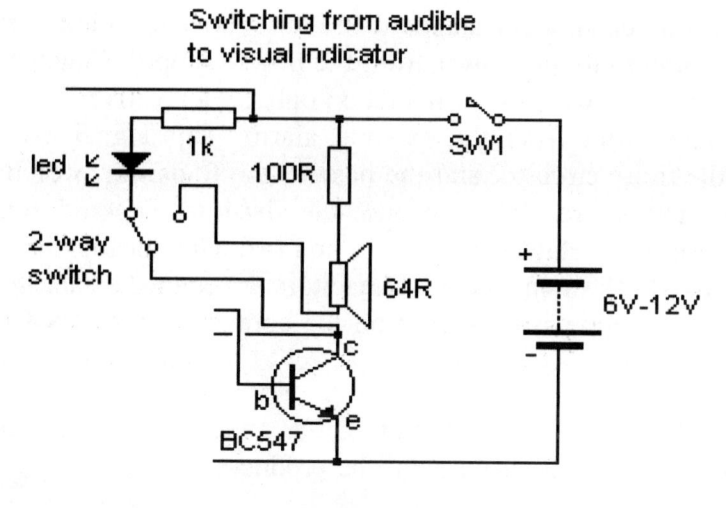

Switching from audible to visual indicator

FIGURE 8.29b

Floodlight switching alarm indicator

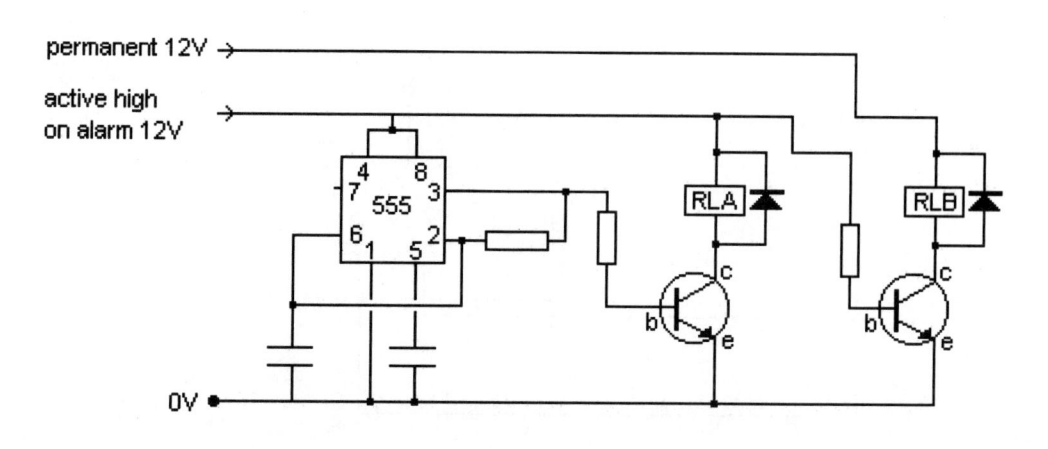

FIGURE 8.30

The design is based around a standard 555 timer device, which is configured as a multivibrator circuit. Power from the power supply is switched by the contacts of relay B, which is switched on only by a positive signal from the alarm-activated wire from the existing alarm. This signal also supplies power to the timer circuitry and the base of the transistor, which switches on relay A. The net result is that once the alarm is triggered, relay A and relay B close, with relay A pulsing on and off. This means that the lamps or indicators will flash on and off. Capacitors are connected across the relay points, because suppressing sparks on the contact points may be required (see Figure 8.30a).

If lamps are wired as shown in Figure 8.30b, using two-way relay switching, then a very eye-catching display will be produced.

Because of the danger of the power supply being damaged by accident or design, it may be wise to replace the floodlights with an array of low-voltage

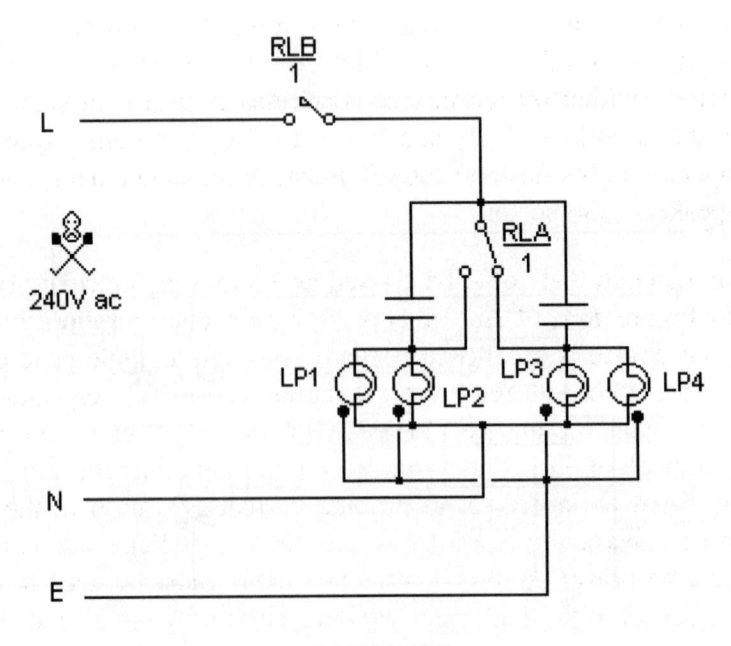

FIGURE 8.30a

```
                LP1   ▫ ▫   LP2
                LP3   ▫ ▫   LP4

                   building
```

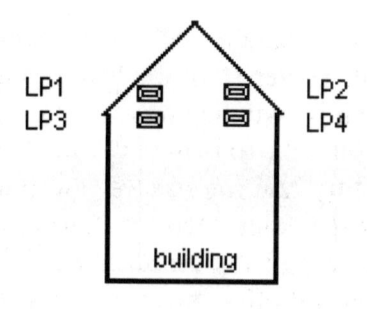

FIGURE 8.30b

strobe lights, which, along with the rest of this circuit, can be driven by a battery backup system.

HUM SENSOR SWITCH, THROUGH-GLASS

This design enables someone on one side of a window to turn on just about anything on the other side of the window that is connected to the switching relay contacts provided by the circuit. This idea is useful in shop window displays, so that a child—of any age (!)—can make a windmill spin around, a train set operate, lights flash on and off, music or announcements come from overhead speakers, and so on.

A circuit for a switch that could be classed as a hum sensor/proximity switch is shown in Figure 8.31. Using a very high-gain configuration with three transistors, the device is able to be triggered through a single pane of glass, or even through double-glazed panels if the air gap is not very large. Extra components are added to ensure a clean switch-on, and a relay is provided to enable the control of larger current. It is important that the high-voltage capacitor be fitted as indicated, to provide correct operation of the circuit, which must be constructed carefully so that the low-voltage side is properly isolated from the power supply. A screened cable must be used to connect between the sensor plate and the switching circuit. A simplified diagram, Figure 8.31a, shows how the hum sensor switch is commissioned.

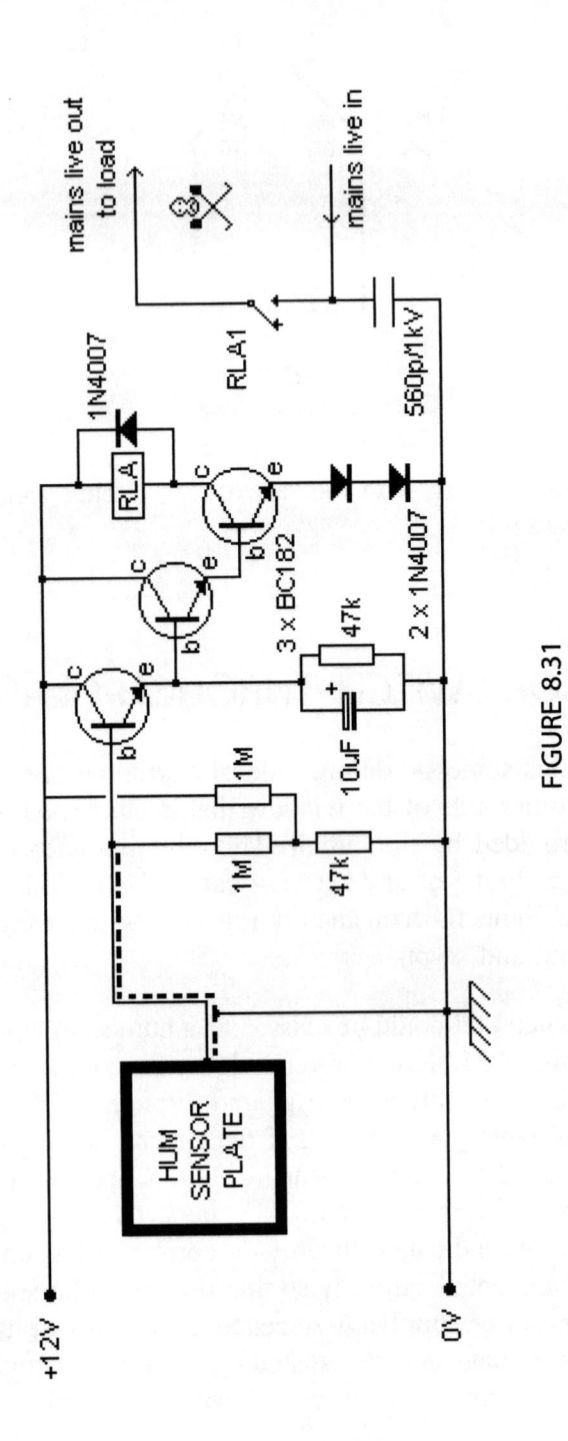

FIGURE 8.31

FIGURE 8.31a

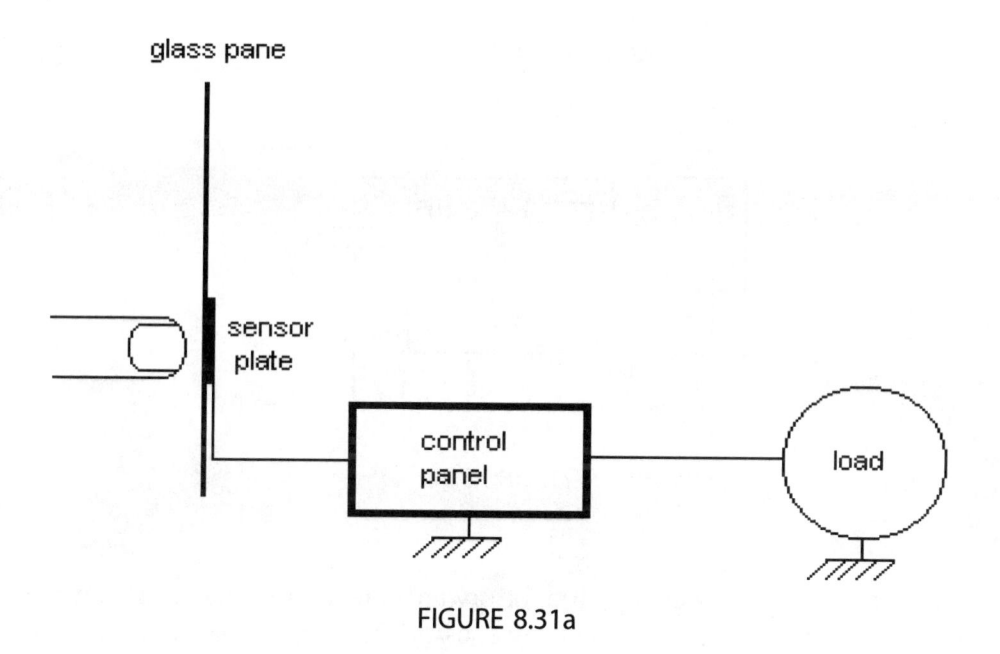

The design may be used in a low-security system for some kinds of switching, although a power supply failure well may render the device unusable. One domestic use, however, could be to switch on internal lighting of a dwelling before entering it, if a particular person is concerned that someone may be inside his or her home waiting in darkness. Using a through-glass switch means that weatherproof switches or associated wiring to an external switch no longer are required.

INDUCTION TRANSMITTER AND RECEIVER: AN ELECTRONIC KEY

A simple induction transmitter and receiver system is shown in Figure 8.32 and Figure 8.33. The designs can be used as part of a security system as a hidden lock, or for arming and disarming systems.

Induction "transmitter" key

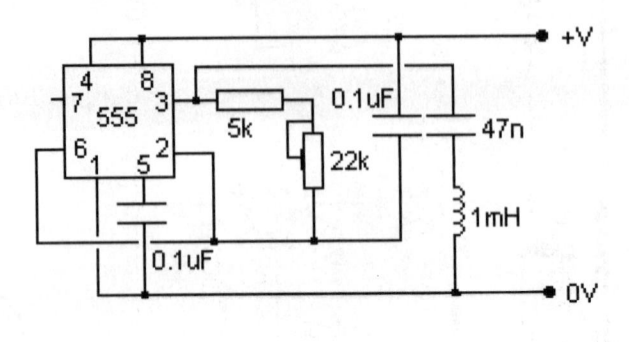

FIGURE 8.32

The operation of the system is straightforward. The transmitter section, which can be built into an old, discarded key-fob-style transmitter, is an equal mark-to-space ratio oscillator based on a 555 timer IC. The output from the oscillator, instead of being connected to a loudspeaker as is the case in many designs shown elsewhere in this book, this time is connected to an inductor. This inductor generates a weak field at the frequency to which the oscillator is tuned.

The receiver section is based on the 567 tone decoder IC, which is shown in Figure 8.33a. This device often is used in an application where it is necessary to decode a tone, such as in RTTY and similar tone-encoded transmissions, so as to obtain an output that then may be processed. The IC also is often used in remote-control applications such as toy cars, where tones are filtered out of the carrier wave so they also may be processed to become control signals to turn the wheels of the car. The signal from the inductance transmitter is picked up by an identical inductor at the "front end" of the receiver and amplified by a single-transistor preamplifier. The pair of back-to-back diodes prevent the input signal from exceeding 0.7V—a useful feature when dealing with inductors that exhibit the property of producing large back-emf spikes when their field collapses. The timing, or tuning, of the 567 IC is performed by connecting components between pins 5 and 6, noting that the tuning pot preferably is of a multiturn variety, to aid accurate tuning. The tuning pot on the trans-

MISCELLANEOUS DESIGNS 159

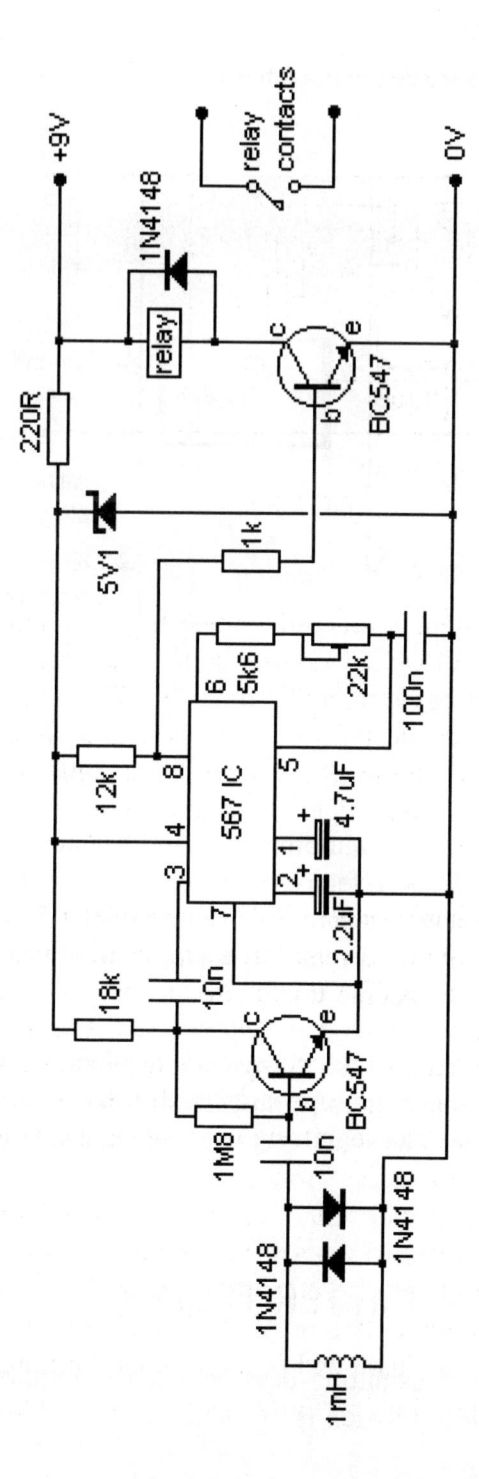

FIGURE 8.33

Electronic key using the 567 tone decoder

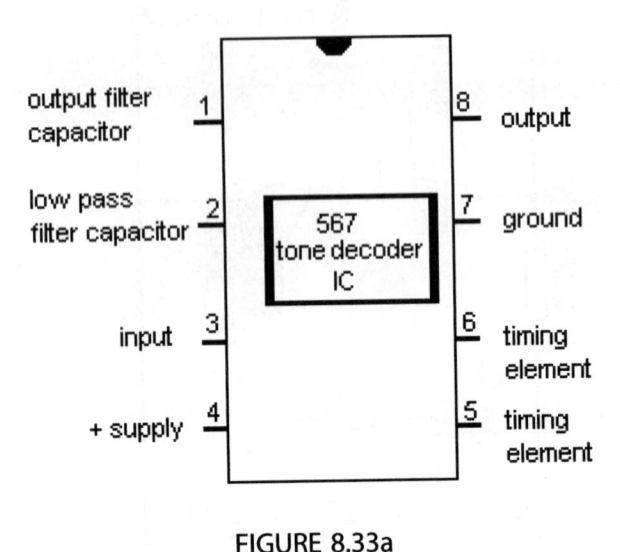

- Pin 1: output filter capacitor
- Pin 2: low pass filter capacitor
- Pin 3: input
- Pin 4: + supply
- Pin 5: timing element
- Pin 6: timing element
- Pin 7: ground
- Pin 8: output

(567 tone decoder IC)

FIGURE 8.33a

mitter section also should be a multiturn pot, and because it is being used as a portable device, should be effectively sealed to prevent the frequency of transmissions being knocked around. The output of the 567 switches a transistor, which in turn switches a relay. The circuit is not self-latching, but a self-latching relay may overcome this.

The output of the receiver could be connected to a solenoid type of lock, which could be used to secure doors or cupboards, with extra transmitter units being produced very easily for other legitimate users of the protected area.

INFRARED TRANSMITTER TESTER

Testing of an infrared transmitting devices often is required to see if it is working. Because the transmission is invisible, it becomes an arduous task to decide whether the transmitter is transmitting or the receiver is not functioning correctly. Sometimes the tone-generating circuit can be heard to work by placing the keyed-on transmitter very close to an off-tuned long-wave radio receiver, where a train of clicks can be heard whenever the transmitter is operated. To actually test infrared transmission, however, a known-to-be-working infrared receiver is required.

Whether you want to test a television remote-control transmitter, infrared night-vision LEDs, or an infrared alarm key fob, the simple circuit shown in Figure 8.34 is an invaluable tool in any kit. The only considerations when building or using the tester are to position the LED and receiver away from each other so as to prevent any feedback, and to remember that the receiving range is only around 20cm (centimeters) or so, depending on the transmission LED power and the receiving diode that is chosen.

Infrared Tester

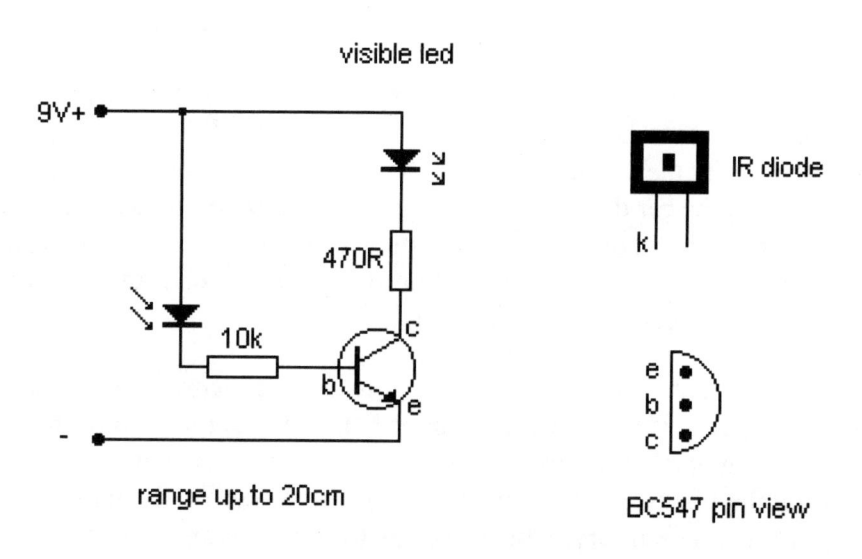

FIGURE 8.34

KEY SENTINEL

The key sentinel can be used either in a domestic situation or as part of any other system. The concept is straightforward: If several keys or bunches of keys are hung on a key rack, this design utilizes eye-catching flashing LEDs to show if any keys are missing, so the situation can be checked by just a quick glance. This system comes into its own if the monitoring LEDs are on a remote board, for example, the monitoring panel could be in an office in a hotel, enabling a supervisor to direct cleaners to an unoccupied room. A security guard could see if all keys had been returned to the key safe a few minutes before checkout time, enabling him to catch the person taking a key outside—perhaps, to be copied.

The circuit could be modified so that if a key or a certain pattern of keys was taken off their hangers, a silent alarm could be triggered. The capability to monitor certain combinations of keys or events could be achieved by using AND gates, so for example, if the safe key AND the emergency rear-exit door is open AND the van keys are missing, an alarm may be initiated.

The only drawback to this system is that wrong keys may be hung in place of the correct ones; however, if personnel are not aware of the system or that the system is being monitored, this will not present a problem.

When the key sentinel is used within a domestic situation, it is useful to check at a glance to see if a bunch of keys are in place; we all know what it's like to have retired for the evening and then have to get out of bed to see if the car keys still are in the car door, or check to see if the garage door keys have been put in their place in readiness for our partner's making an early start while we still slumber. A flashing LED is far easier to see than bunches of keys hanging off a single nail stuck in the front door.

An idea for constructing the switching is to use tamper switches with long arms (see Figure 8.35). These arms can be bent and fashioned to form a shape suitable for hanging key rings with ease. I suggest that soft heatshrink sleeving be used on the hooks, so that any sharp edge or corner on the tamper switch arm does not cause injury. The fascia of the key sentinel can be made of hardboard or thin plywood, allowing sufficient horizontal movement of the switch arms by cutting slots in the board.

MISCELLANEOUS DESIGNS 163

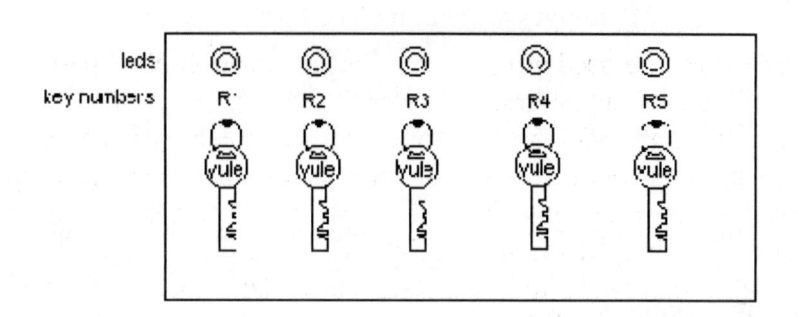

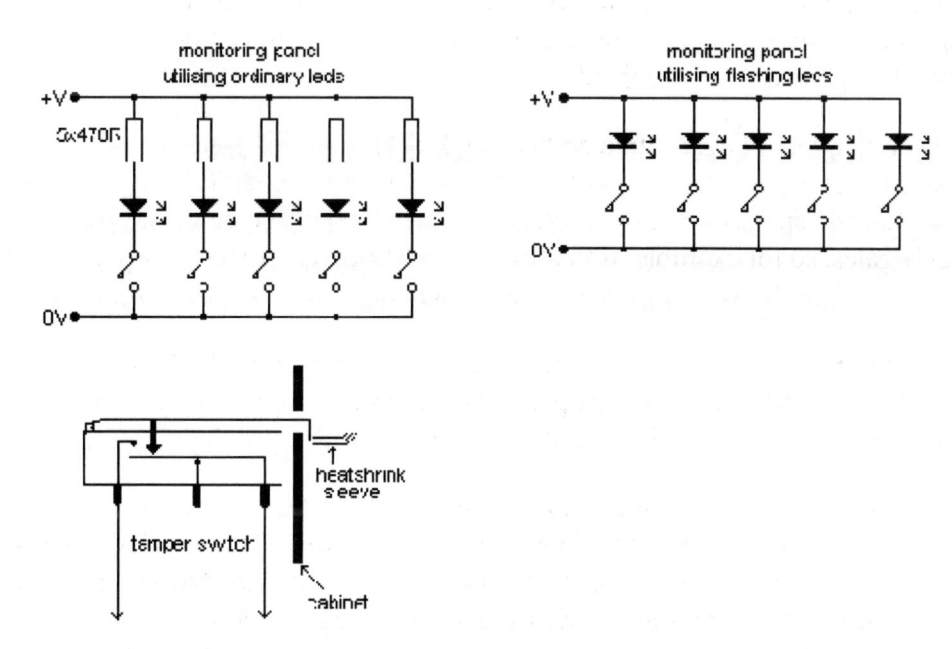

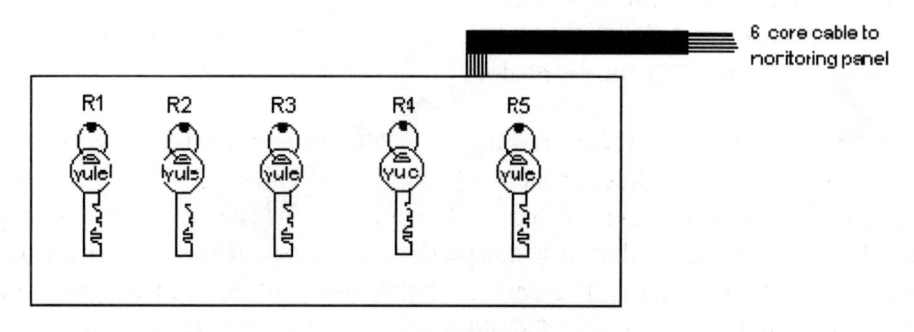

FIGURE 8.35

The monitoring circuit may consist of special flashing LEDs, some of which will work on a range of voltages without the need for series resistors. If there is a need to economize, then standard LEDs with 470R series resistors may be used. Because the unit will be used around the clock, I suggest that the circuit be powered from a low-voltage power supply, at around 6V to 12V DC output.

If the key sentinel is to be monitored at a remote point, then a connection using multicored cable must be run from the key safe to the monitoring point, unless a multiplex circuit is employed.

LIE DETECTOR/DAMP WALL DETECTOR

Figure 8.36 shows a circuit that can double as a lie detector and to hunt out dampness in a dwelling's internal walls. The circuit is a simple three-transistor, high-gain amplifier with a variable resistor that acts as a sensitivity

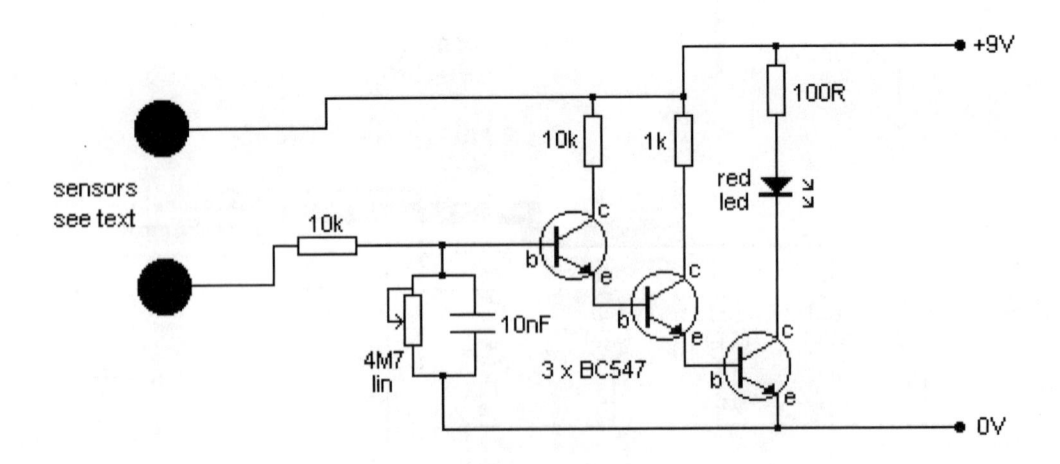

FIGURE 8.36

control. The 10k resistor at the input is simply to prevent burnout of the transistors if the probe lead touches the positive battery terminal.

If the circuit is commissioned as a lie detector, you first of all must recognize that this concept is for "fun only," and therefore should not be used for any other purpose. The idea is an old one: believing that as someone who is not telling the truth will perspire more than if he or she was telling the truth. With this concept in mind, form two probes by soldering the input wires to copper coins, previously cleaned and polished with wire wool.

Place the probes a few centimeters away from each other onto the palm of the person being interrogated, and fix the probes firmly with adhesive medical tape. Then connect power, which for safety reasons must be supplied only from a battery, to the circuit. Adjust the sensitivity control after a minute or two, to allow the victim to settle down, and then ask questions that can be answered truthfully and without causing stress. Next, adjust the sensitivity control until the LED only just extinguishes. When you now ask the person a question that causes stress either to answer or contemplate, the skin resistance lowers as perspiration forms. The decrease in skin resistance causes an increase in voltage to the base of the first transistor, which switches the amplifier and LED full on.

Damp Detector Assembly

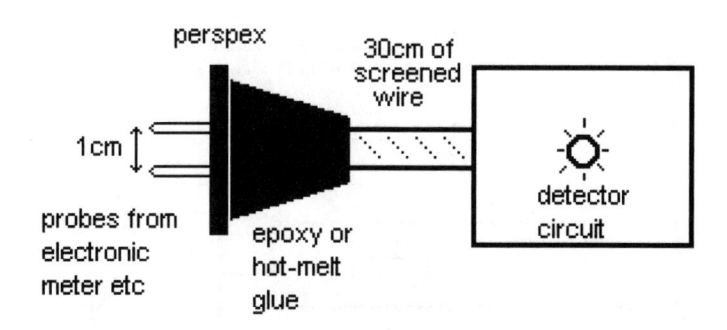

FIGURE 8.36a

If the circuit is commissioned as a damp detector, the copper coin probes are replaced with an assembly, as shown in Figure 8.36a. You can build the whole device into a box, but if hard-to-get places of a wall need to be reached, then you can use a short length of screened cable to connect the circuit and probe assembly. A pair of pointed instrument probes are ideal for poking through wallpaper that may be hiding metal foil, which often is used as a means to temporarily prevent dampness from showing through.

LOOP ALARM SYSTEM

A loop alarm system has many uses due to inherent flexibility. It can be used inside doors, walls, floors, and ceilings to give an indication of a hole-in-a-wall-style of attack. Figure 8.37 shows an example of a loop alarm that is used

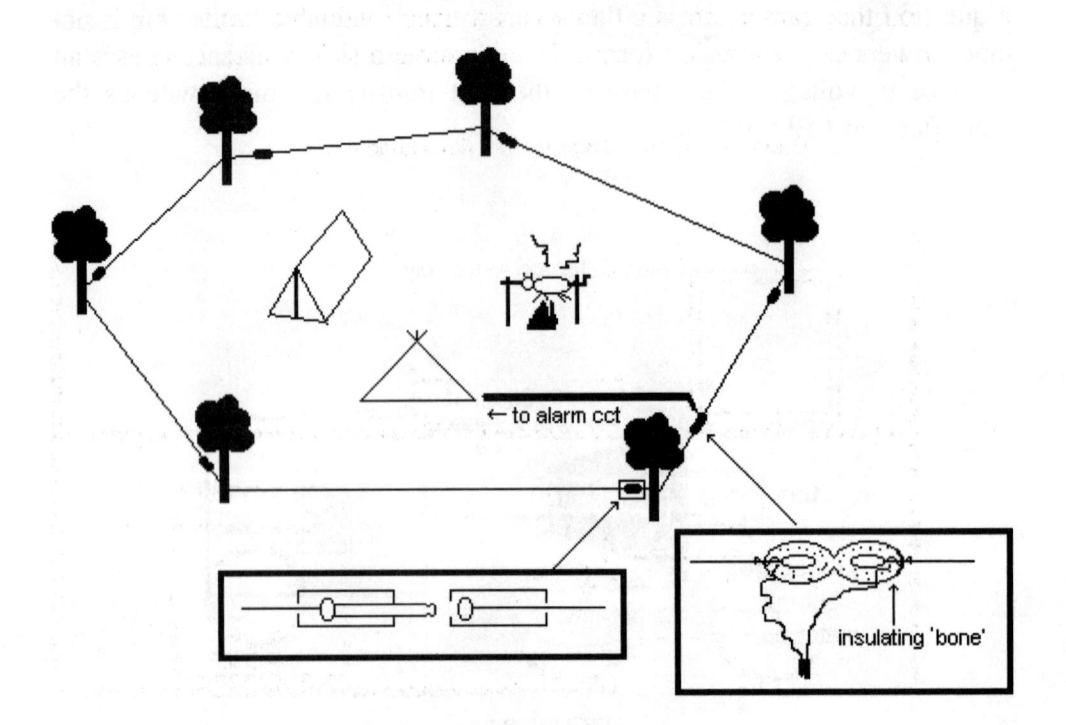

FIGURE 8.37 Camp Alarm System

around a camp perimeter to give a warning of persons entering the camp area. The loop is set up as a trip wire so that, unless the intruder is crawling along on the floor, a wire made from strong, insulated multistrand material is pulled apart at set break points by the intruder. You make the break points simply by using plugs and in-line sockets.

Considerations must be made concerning the length of each run and the weight of the trip wire, because if the length is too long or the wire too heavy, any attempt to pull up the center sag simply will pull the break point apart. To this end, it may be pertinent to introduce some kind of clotheslines made from handy twigs or small branches. Some type of insulator, such as a plastic dog bone, will be required at the point where the two free ends of the loop go to feed the alarm circuit. If time does not allow as complex a setup as described, obtain a large reel of very fine enamelled copper wire and use this as a perimeter trip wire, although this necessitates rewiring the loop each time it is broken.

Figure 8.37a shows where a versatile wire loop system may be implemented in a workshop, garage, or retail premises to guard valuable items. If it is not feasible to pass the plug and wire through some articles, such as television

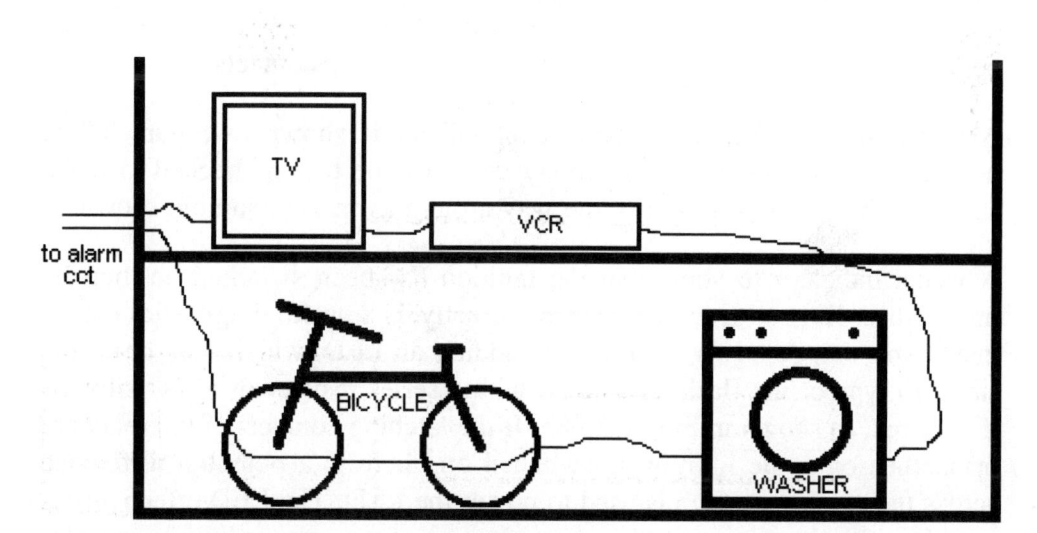

Example of protecting property with a Loop Alarm

FIGURE 8.37a

sets, stick-on brackets with loops can be used. A key switch is used to arm or disarm the loop alarm while items are moved to or from shelving.

LOW-COST VEHICLE IMMOBILIZER

A simple but efficient method of immobilizing a vehicle is to have a broken connection in the supply line of any electrically powered part of the engine, which could be an electrically driven fuel pump, ignition wiring, or starting motor solenoid.

To reconnect the broken wire, use a set of relay contacts, as shown in Figure 8.38. To energize the relay, take a line from some point that becomes positive only after the ignition switch is turned. Because there is a set of relay contacts in series with the relay coil, you must now energize the relay coil so that the latching contacts can operate. Do this by utilizing an existing apparatus switch, a hidden switch, or a key switch. Once the relay is energized, the first pair of relay contacts latch the relay on. A further pair of relay contacts, which have been wired in parallel across the broken wire, also close, so "mending" the break. Two diodes are used in this circuit: the first connected across the relay coil so as to prevent back emf whenever the coil is switched on or off, and the other diode connected to prevent the apparatus, such as a rear window, from being continuously powered up via the latching relay contacts.

A wiring diagram of the low-cost immobilizer is shown in Figure 8.38a; however, I recommend that any relay used first be bench-checked to make sure that the contacts are the same way around as in the example shown.

A visual indicator to show that the ignition has been switched on, but the immobilizer has not been deactivated correctly, is a useful diagnostics instrument. This can be easily achieved by adding an LED, which may be of the standard type or the flashing variety, with a series resistor of a few kilohms if required, as shown in Figure 8.38b. If the latching contacts of the relay are open, then once the ignition is switched on, there is a potential difference across the contacts, which is used to power the LED. The LED effectively is in series with the relay coil; therefore, the size of the current-limiting resistor depends on the resistance of the relay coil used.

MISCELLANEOUS DESIGNS   169

Low-cost vehicle immobiliser

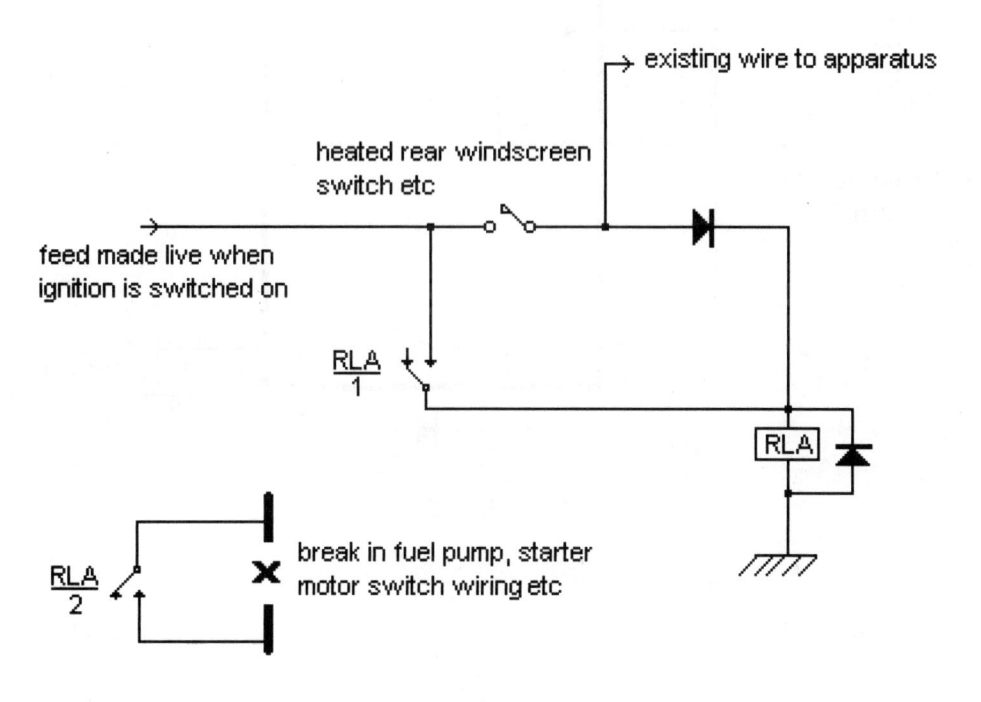

FIGURE 8.38

POWER SUPPLY-ON ALARM

You well may be wondering why an alarm is needed to indicate that the power supply is on, when all you have to do is hear a television set, switch a light on, and so on! This circuit was designed for safety in the kitchen, because many electric-cooker hot plates can be fully turned on without the user's being aware that this is the case; where the older ring style gave a healthy red glow, many modern hot plates look the same whether switched on or off. Another use for this design is to remind the person using the cooker that it is switched on, should that person be busying himself or herself outside the kitchen area.

170 APPLIED SECURITY DEVICES AND CIRCUITS

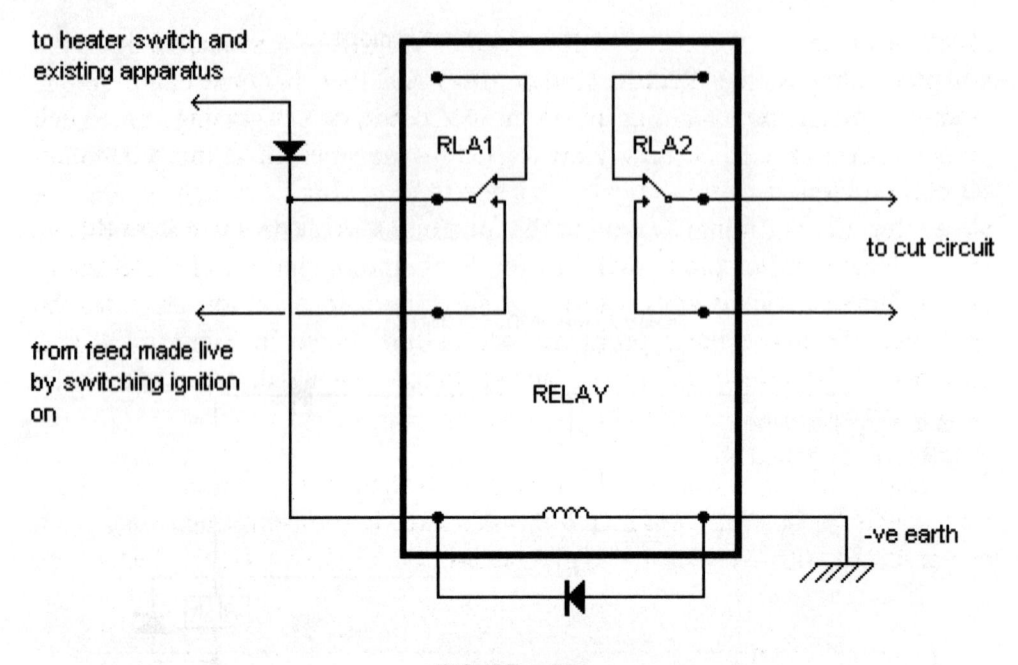

to heater switch and existing apparatus

RLA1 RLA2

to cut circuit

from feed made live by switching ignition on

RELAY

-ve earth

FIGURE 8.38a

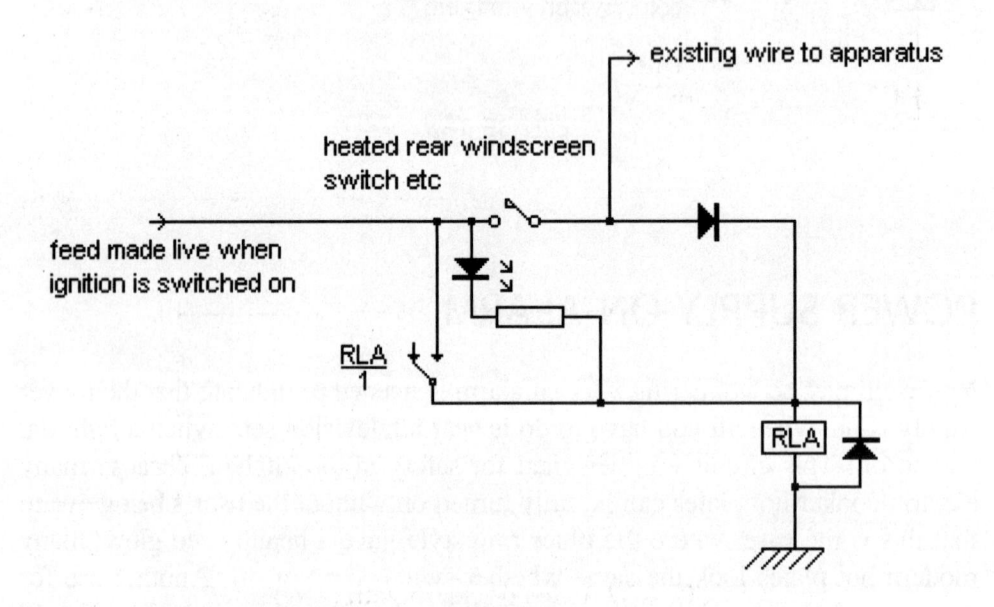

existing wire to apparatus

heated rear windscreen switch etc

feed made live when ignition is switched on

RLA 1

RLA

FIGURE 8.38b

MISCELLANEOUS DESIGNS

Many of us have put a piece of food under the grill or a saucepan upon the hot plate (that is, the electric cooker ring) and then become engrossed in reading a magazine, catching up on the TV news, or performing some such joyous chore, only to be rudely awakened by the smell of burning. Another serious problem we often may experience is forgetting to switch off the hot plate after use. We hungrily devour the food before rushing to work, but leave the appliance and/or plate switched on, hence producing a hefty electricity bill, a fire risk, or possibly causing a nasty burn to a person cleaning the appliance. To avoid these problems, the circuit shown in Figure 8.39 was designed to be connected to the power supply whenever the appliance is switched on at the wall switch.

The simple circuit generates a short, loud beep about every five seconds; if the user is hard of hearing, an LED with associated current-limiting resistor can be paralleled with the sounder to give an additional visual warning. Although

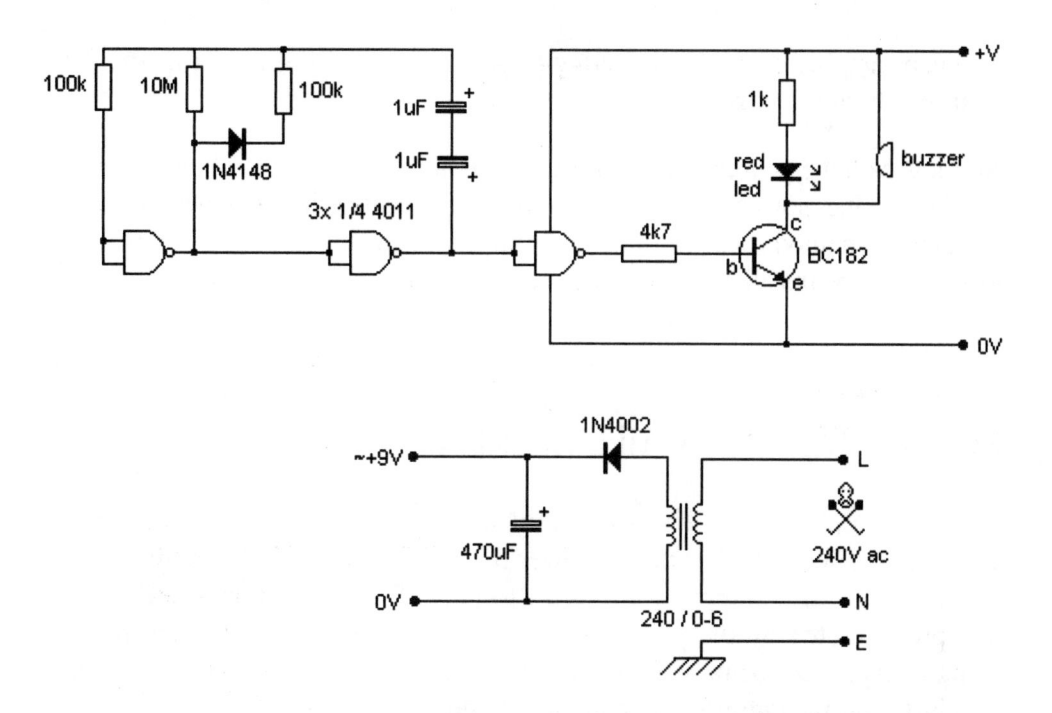

FIGURE 8.39 Mains-on alarm with bleeping Audible and Flashing Visual Indicator.

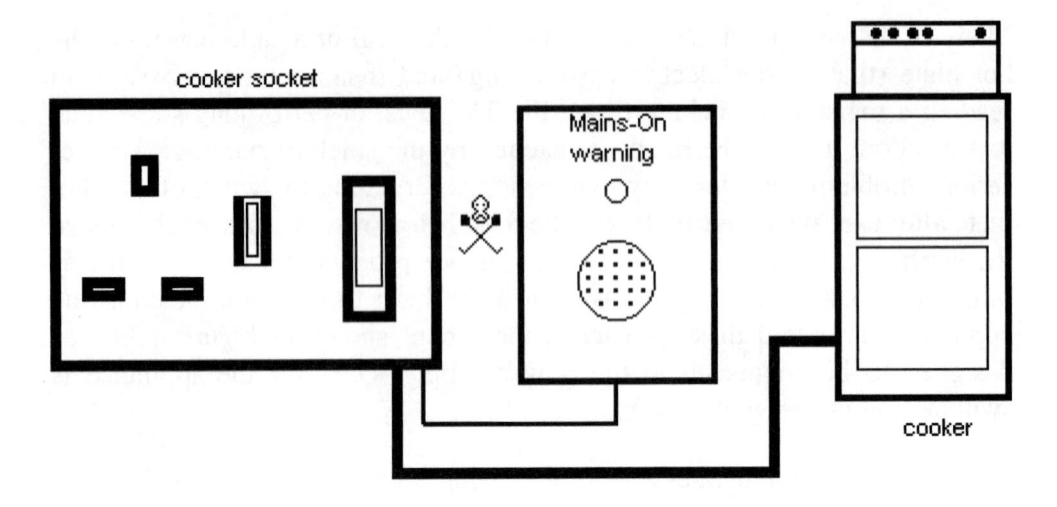

FIGURE 8.39a Installation of Mains-on Warning Unit

the beep may be somewhat irritating to some users, still, it performs the duty required: to act as a reminder.

Any low-voltage source can be used as long as it is ac-powered and operated from the same power switch as the appliance. Figure 8.39a shows a typical installation method; bear in mind that the power switch should be completely isolated before installation work begins, due to the risk of electric shock.

POWER SUPPLY FAILURE ALARM

A power supply failure alarm is useful for alerting a person that the power supply has failed, whether due to power lines being knocked down by a storm, vandalism, or an attempt by attackers to cut off the supply before breaking into premises fed by the power line. If a large amount of expensive food is being refrigerated, or maybe drugs or plasma, it is imperative that the power supply to the refrigerator be maintained and monitored.

Two designs for such a monitor are shown here. Figure 8.40 is a nonlatching unit that will sound only during the period of time when the power is absent

MISCELLANEOUS DESIGNS | 173

Mains Supply Failure Alarm

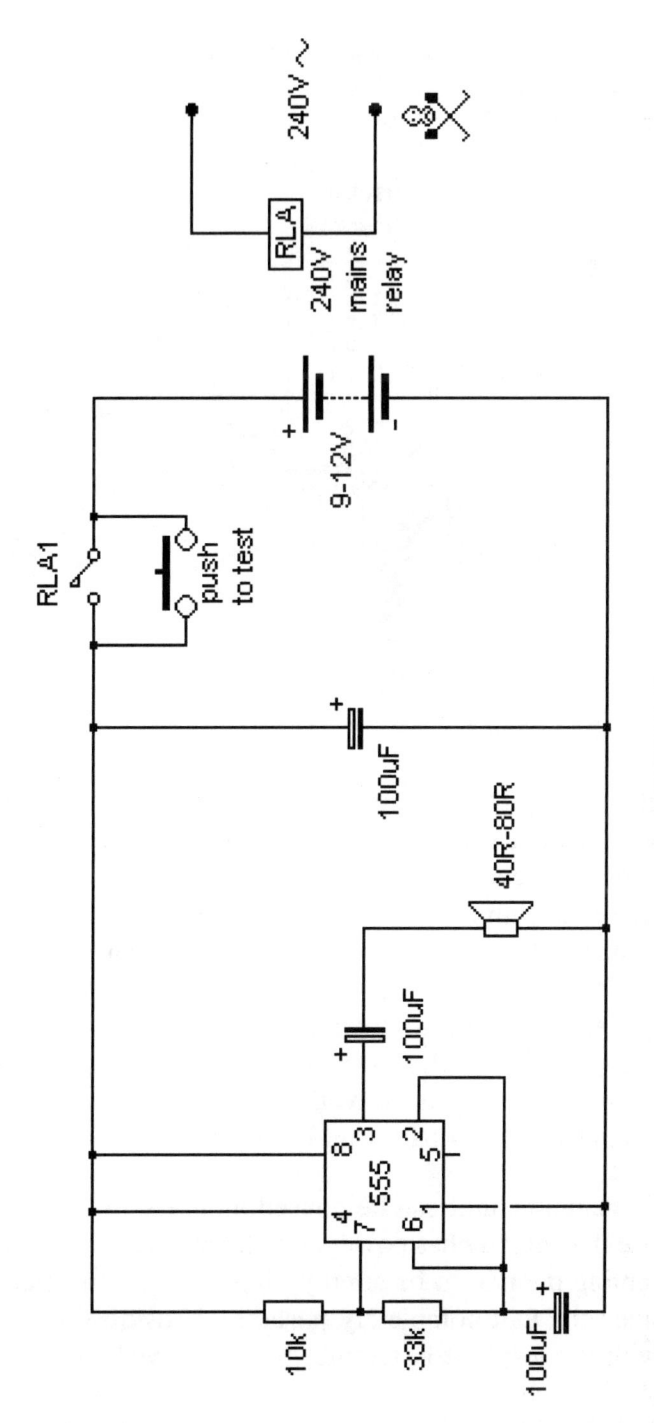

FIGURE 8.40

Main Supply Alarm Unit

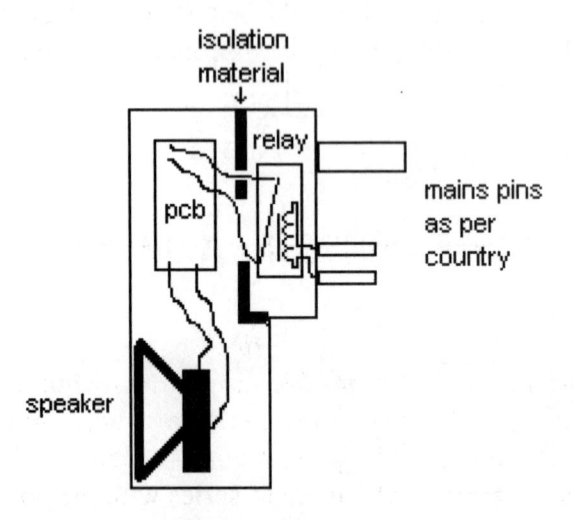

FIGURE 8.40a

and then silences once the power is switched back on. The circuit uses a 555 oscillator with audible output. A relay is connected in series with the power supply, where relay contacts that stay open while the relay is energized are connected in series with the supply to the oscillator. If the power supply fails, the relay contacts close, powering up the alarm circuit. A test button, which is connected in parallel with the relay contacts, is pressed closed to periodically test the alarm circuit to make sure that the battery is functioning correctly. A safety note: As with all circuits connected to the power supply, make sure that the power supply is correctly isolated from the low-voltage circuit and that no exposed metal parts can become live.

If required, the whole unit can be housed in an empty power-supply case—a plastic case that already has power connector pins on it, as shown in Figure 8.40a—enabling the unit to be simply plugged in to any electrical socket in order to make the unit completely portable. If using this method, you can dispense with the test-button feature, because checking to see if the alarm is working satisfactorily requires only periodically unplugging it. If the ac wiring is in the same enclosure, avoid excess copper being exposed and fit an isolation barrier fabricated from a sheet of plastic, for example, between the low- and high-voltage sections. ALWAYS unplug or disconnect from the power supply before connecting or disconnecting a battery to avoid an electric shock hazard.

POWER SUPPLY FAILURE ALARM WITH LATCH

Figure 8.41 illustrates a power supply alarm system with a latching action. If the power supply fails but comes back on again, unlike the previous example, this latch continues to give an audible alarm. This is because the second set of relay contacts, which are in series with the power supply relay coil, are closed only by energizing the relay, which initially can be performed only by using a *set* switch. The set switch is depressed, causing the relay coil to energize and so close the contacts RLA2. RLA1 contacts now are open. The alarm sounds only if the test button is closed, or if there is a power supply failure, in which case the RLA2 contacts open as the relay coil de-energizes, and contacts RLA1 close.

METAL DETECTOR/"HIDDEN TREASURE LOCATOR"

This circuit is based on the superheterodyne principle used in basic radio receivers. The superheterodyne system works thus. Two L (coil) and C (capacitor) oscillators, which work on the same frequency, have their outputs both sent to a mixer circuit. One of the oscillators will be of fixed frequency, that is, the reference oscillator. The other oscillator, however, will have a coil—the search head or detector coil—whose inductance will be altered by

176 APPLIED SECURITY DEVICES AND CIRCUITS

Mains Supply Failure Alarm with latch

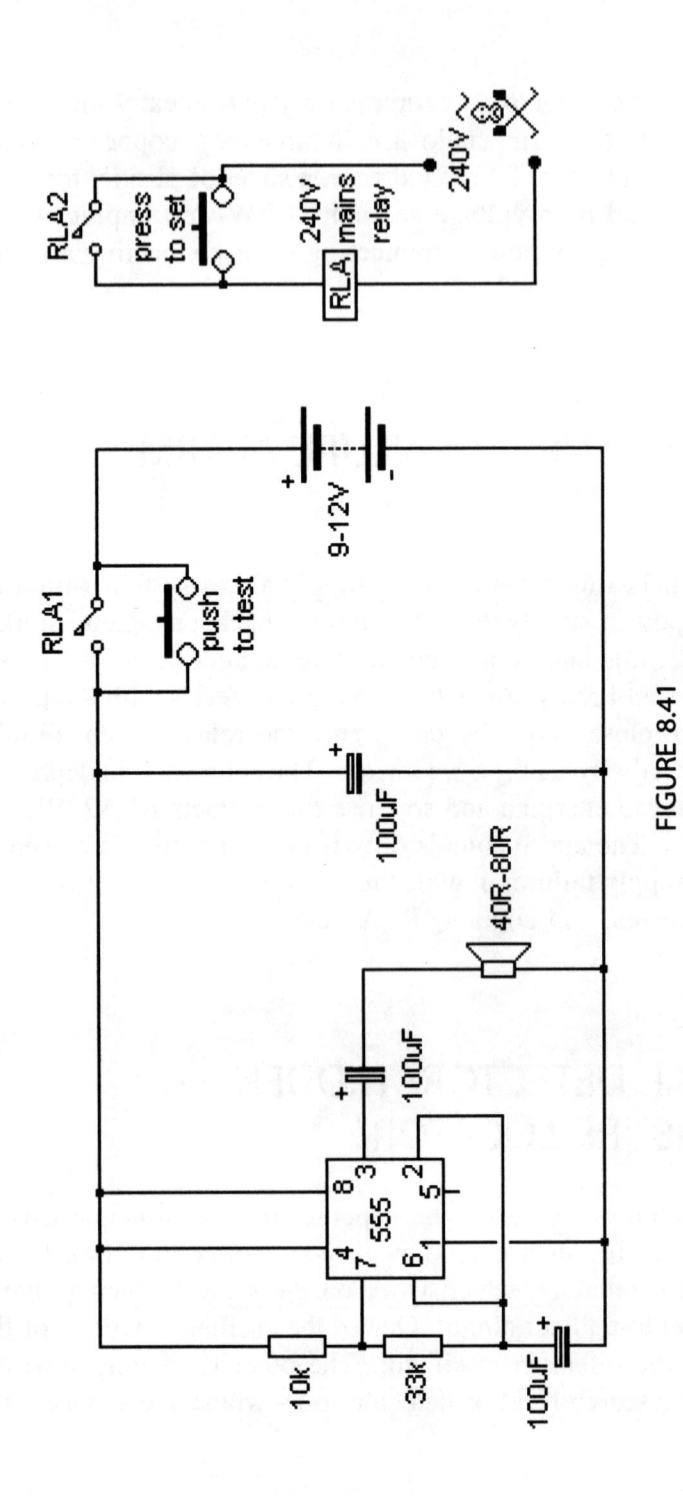

FIGURE 8.41

a piece of metal coming into close proximity. Because the operating frequency of an LC oscillator is dependent on the value of the coil and capacitor, the altering of the inductance by the piece of metal will produce a frequency shift of this oscillator. Refer to Figure 8.42 to see that the mixer circuit will generate many outputs, for example:

f1

f2

f1 + f2

f1 − f2, etc.

If f1 is the reference oscillator and f2 is the detecting oscillator, if, for example, both are set to exactly the same frequency, say 465kHz, then among the outputs from the mixer will be the following:

f1 = 465kHz

f2 = 465kHz

f1 + f2 = 930kHz

f1 − f2 = 0kHz

Now, if a piece of metal comes close to the search coil, upsetting the balance, it may alter the frequency of the detecting oscillator by 1kHz, pulling the frequency of the detecting oscillator down to 464kHz! Then among the outputs from the mixer would be:

f1 = 465kHz

f2 = 464kHz

f1 + f2 = 929kHz

f1 − f2 = 1kHz

The "Superheterodyne" Principle

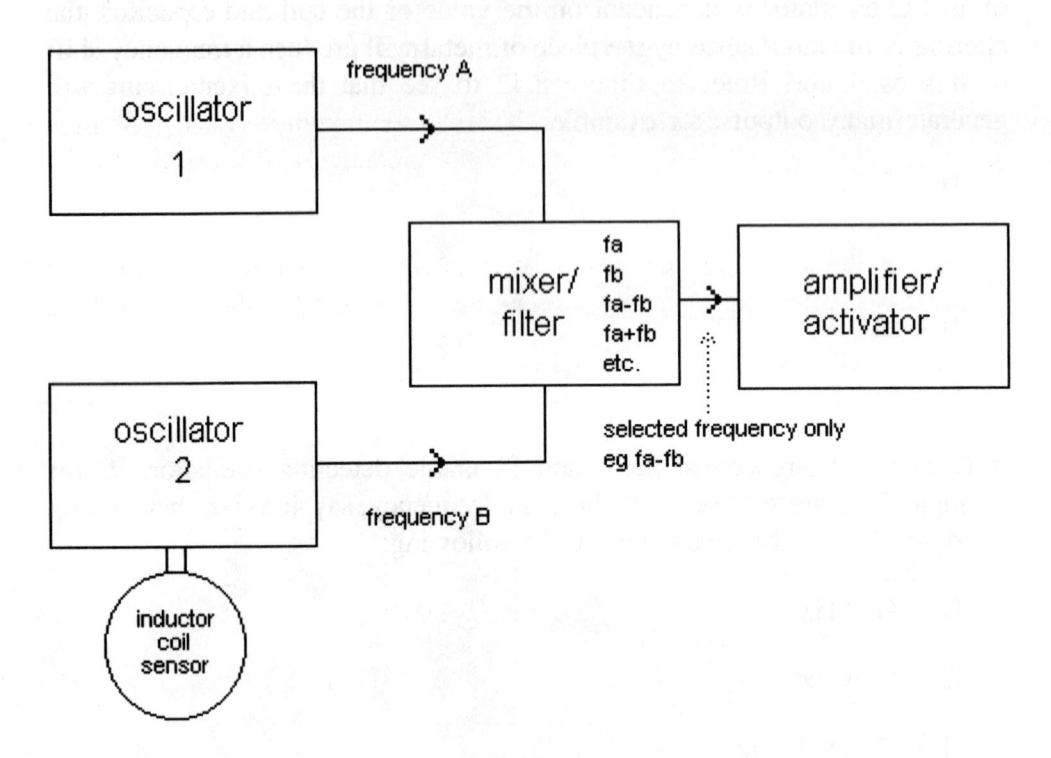

FIGURE 8.42

This 1kHz difference now can be sent to a pair of headphones, an audio amplifier, or some kind of activator or alarm circuit. The mixer usually is followed immediately by a filter that blocks all the unwanted products, possibly simply taking the form of a small-value capacitor that filters out the high frequencies down to ground yet gives sufficiently high impedance to the audio tone produced.

Because the output of the mixer relies on the two oscillators, both these oscillators should be well constructed to prevent frequency drifting. The reference oscillator need not be of an LC design but could be constructed using a crystal oscillator.

A simple metal detector circuit is shown in Figure 8.43. The reference oscillator is made from a circuit containing an IF (intermediate frequency) can from a discarded AM transistor radio. The circuit does not use a mixer as such, because the output from the search oscillator is coupled to the reference oscillator via the 10pF capacitor. This pretty crude output is connected to the audio stage through the detector-style diode and the 4u7 capacitor to the transistor amplifier.

A basic actuator that can be used with a metal detector circuit is shown in Figure 8.44. The preset resistor acts as an attenuator for the input signal,

Basic metal / hidden "treasure" detector

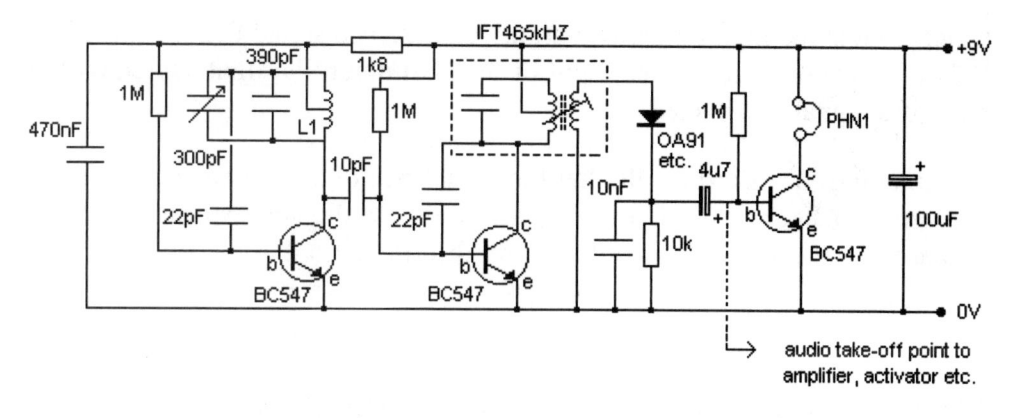

construction of sensor head (L1)

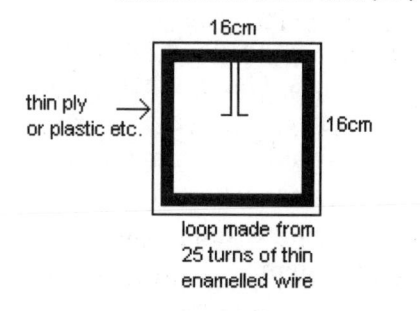

IFT465kHz = IF transformer from old am radio
= constructed as shown
PHN1 = 2k impedance earphone, earpiece etc.

loop made from 25 turns of thin enamelled wire

FIGURE 8.43

Metal detector activator circuit

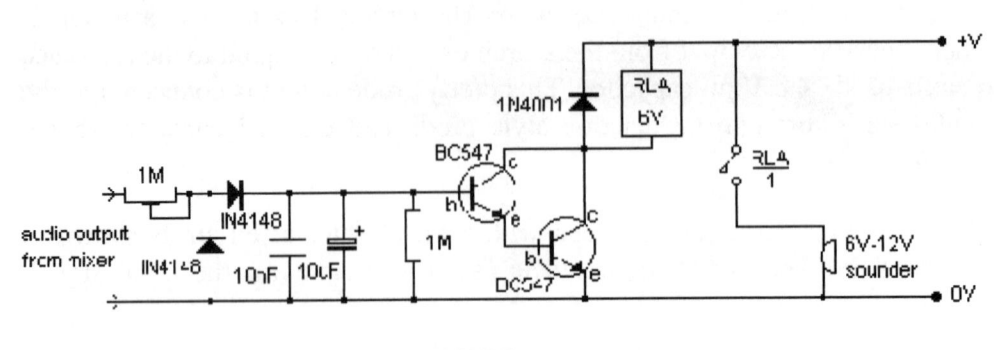

FIGURE 8.44

which then is rectified by the two signal diodes. The signal then is smoothed before being connected to the input of the Darlington Pair that switch on the relay if a signal is detected.

Note that ferrous material will alter the inductance of a coil one way, whereas nonferrous material will alter it the other way, but the net result is the same, because f1 – f2 gives an output with no discrimination!

MULTI-CAMERA SWITCHING CONTROLLER

If more than one surveillance camera is installed at one or more sites, then some method is required to monitor all camera outputs. One method uses a single video signal monitor with a split screen, where each camera has a different code. Another method uses multimonitors, where each camera has a separate monitor. A third method uses a time-lapse system, where the monitor flashes between each surveillance camera, and the resultant mix of images is separated by a control unit in the video-recording system.

A somewhat cheap and cheerful compromise is shown in Figure 8.45 and Figure 8.45a. This design is very basic in operation, where a 555 timer IC with

MISCELLANEOUS DESIGNS

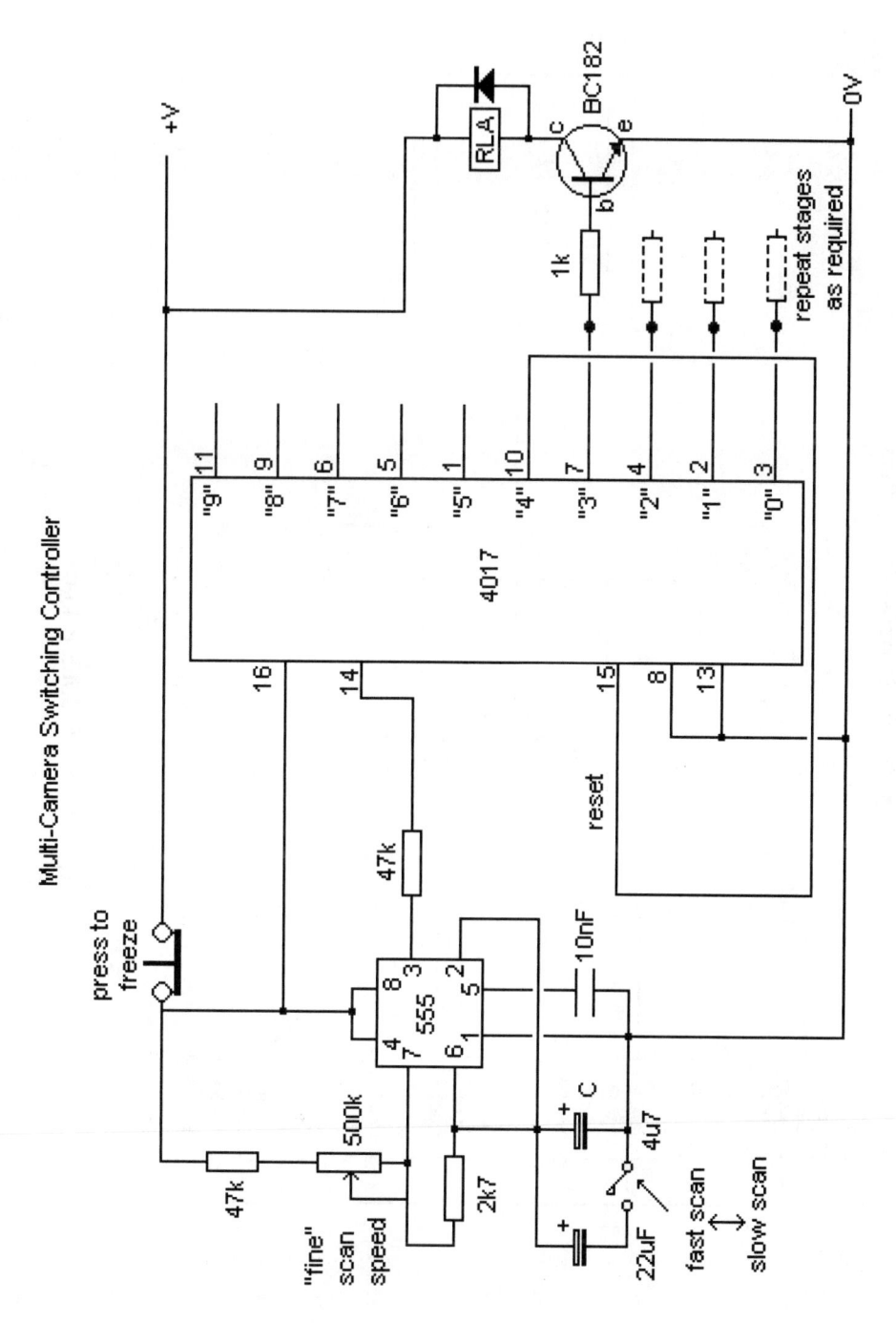

FIGURE 8.45

Multi-Camera Switching Controller

182 APPLIED SECURITY DEVICES AND CIRCUITS

Multi-Camera Switching Wiring Details

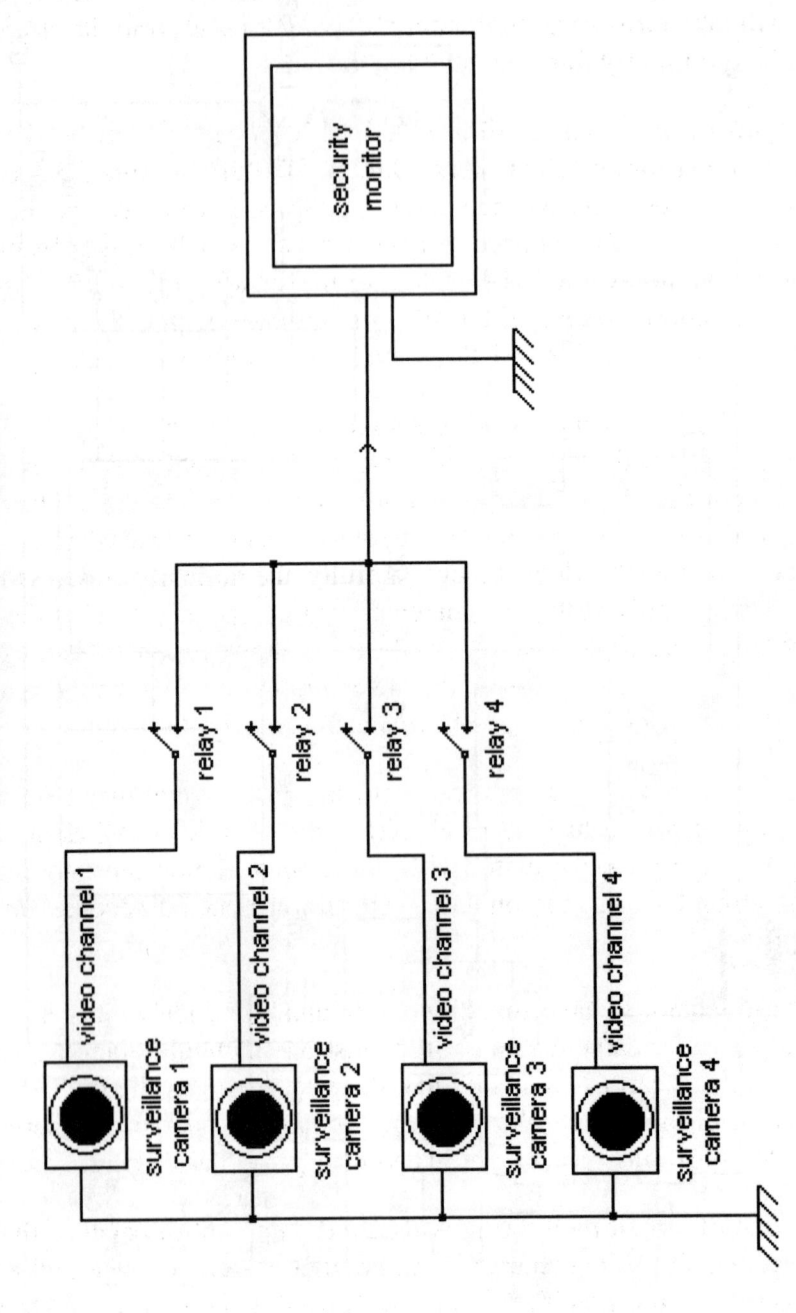

FIGURE 8.45a

varying timed output is connected to a 4017 IC, a divide-by-10 device, whose outputs will take turns to go high after receiving a pulse from the clock, with the preceding output going low, and so on.

Each output—in the example shown, "0," "1," "2," and "3" of the 4017—is taken to a transistor-switched relay. Output "0" must be used, otherwise a blank gap will appear during the scanning process. Each relay-switched contact is taken to each surveillance camera, with all switch outputs being connected to a common cable that is taken to the security monitor/VCR or the like. The next unused output of the 4017, in this case output "4," is connected to the reset pin, so that the switching cycle begins all over again.

By varying the rate of the clocking signal, you can scan through the video channels at any speed required; as you can see by the 555-driven circuit, the timing components C and R can be fully adjusted to suit any requirement. Because it also may be required to freeze the scanning procedure so that a particular scene may be viewed more carefully, the normally closed switch in series with the supply voltage to the clock can be opened, pausing the clock and so stopping the 4017 at any point. Be aware, however, that if a smoothing capacitor still is connected across the clock supply, then where the scanning actually stops will be potluck, as the capacitor slowly discharges!

Although a 555 timer is used in the design, any clock circuit may be used, for example, one made from CMOS NAND gates of a 4011 IC, as shown in Figure 8.45b. In this case, a switch will cut in a set of timing capacitors, giving a rate of around 6 seconds on fast scan and around 60 seconds on slow scanning.

It is a useful feature to have some kind of manual override so it is easy to go back to a particular camera shot or study just one particular danger area. For this purpose, you can use a switch like the one shown in Figure 8.45c. The first position of the switch allows normal automatic operation, whereas the other switch positions pick off the video of a particular camera.

If some kind of circuit monitoring is required on a control panel so that it is easy to see which of the cameras is online, then a small change to the basic circuit, as shown in Figure 8.45d, may be utilized. Each used output from the

184 APPLIED SECURITY DEVICES AND CIRCUITS

Multi-camera switcher using CMOS clock

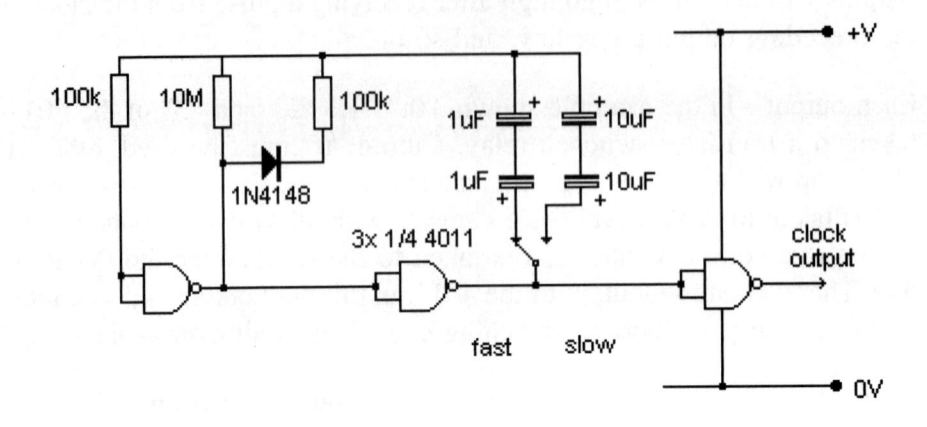

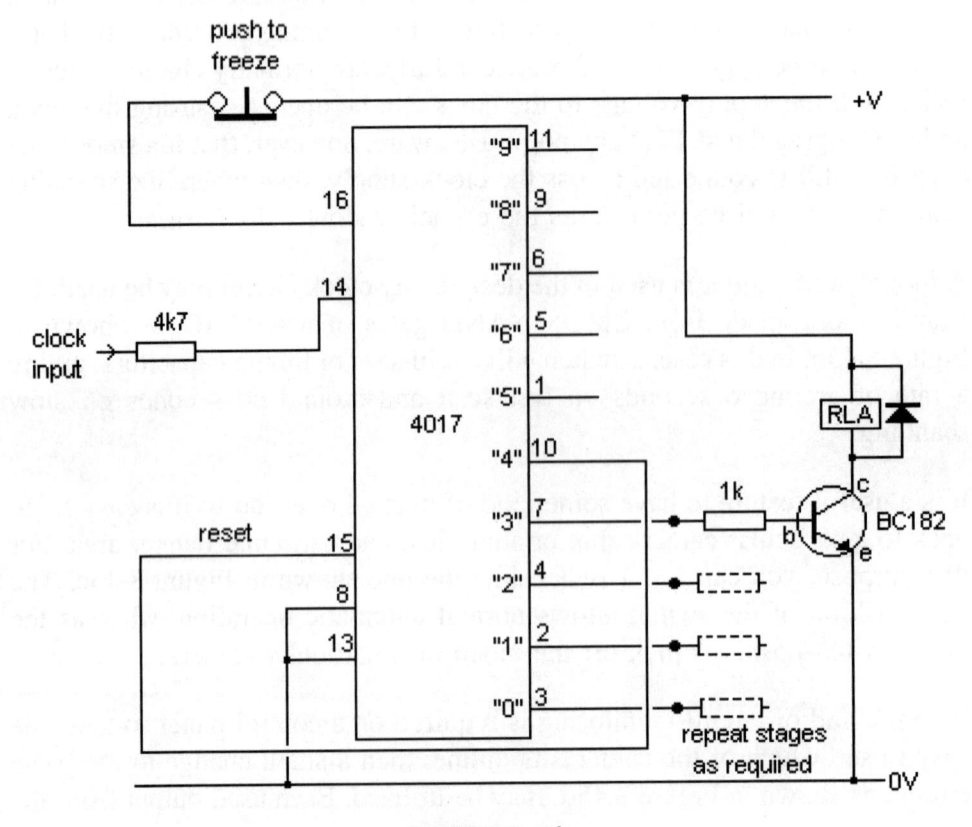

FIGURE 8.45b

MISCELLANEOUS DESIGNS   185

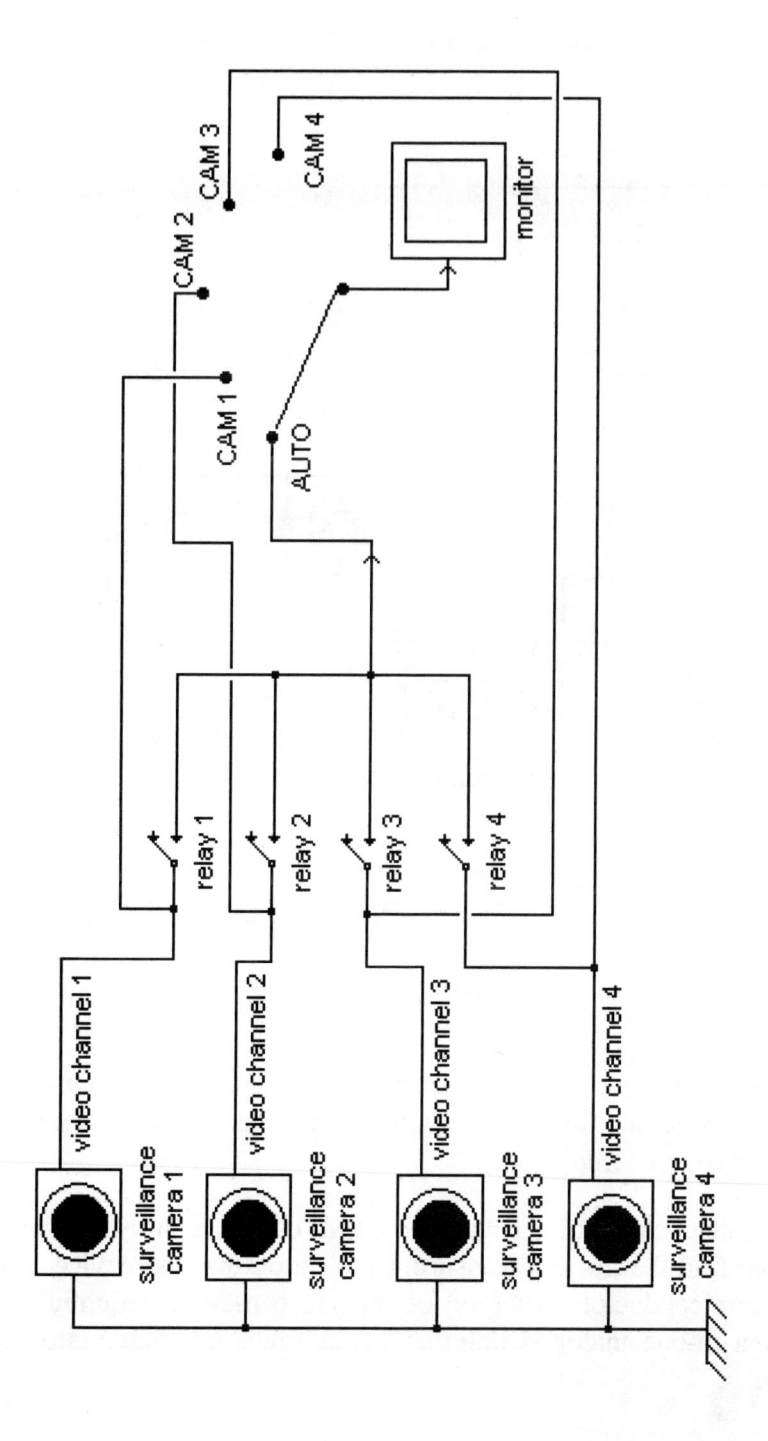

FIGURE 8.445c

Multi-camera switch LED monitoring

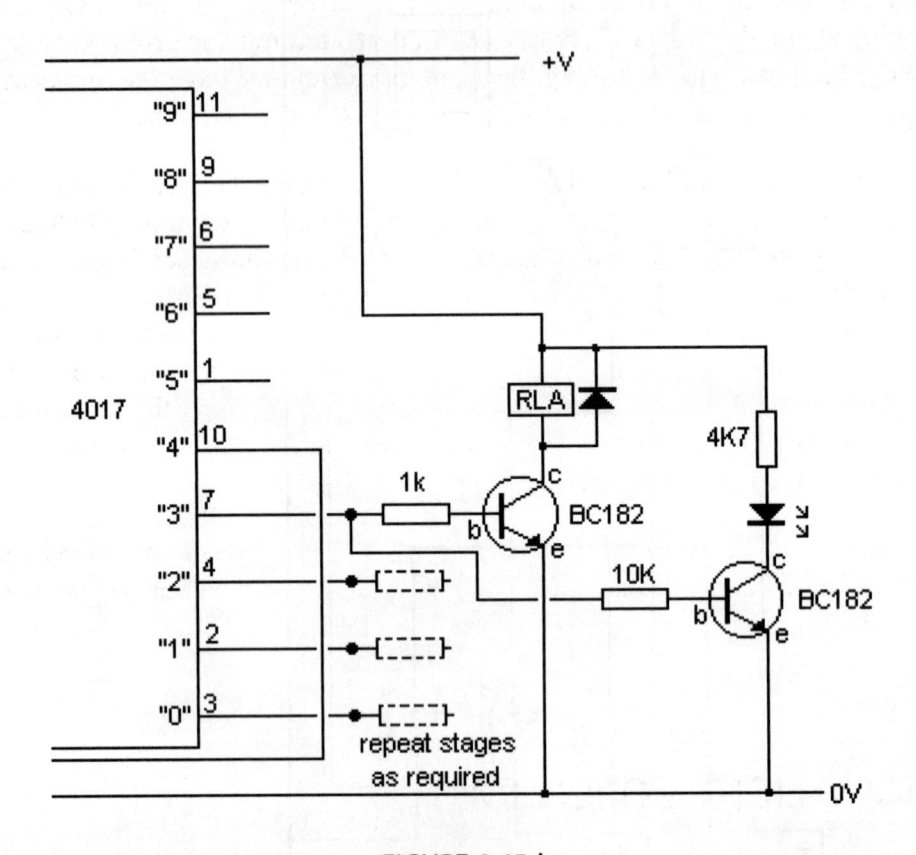

FIGURE 8.45d

4017 is connected to a further transistor that drives an LED mounted on the control panel. Note that this method will not indicate which camera is online if manual override is employed.

If the constant clicking of mechanical relays causes a problem—although some people find the noise a reassuring indication that the device is working—then semiconductor switching of the video may be required; for example, a four-in-one analog switch that can be found integrated into a single IC package.

NICAD BATTERY MONITOR

You can use this design to monitor any type of battery, of any voltage, by making slight alterations to component values. The circuit is useful for monitoring the nicad type of battery, because these batteries have the unpleasant characteristic of just giving out without a great deal of warning.

The circuit shown in Figure 8.46 consists of two parts: The left-hand section is the comparison section and the right-hand portion is the warning indicator, which may be any low-consumption design. A multiturn pot is connected across the supply rails, and is set so that the base of the first transistor is on, which means that the second transistor will be off, and so no power will be connected to the alarm indicator circuit. When the supply line drops to a predetermined level, which has been set by the multiturn pot, the zener diode stops conducting; the first transistor switches off, and so the second transistor switches on, so powering the alarm indicator.

The circuit is suited equally to monitoring the progression of a lead/acid battery being charged, where the LEDs will flash until the battery has been charged up past the "danger, low battery" point.

NIGHT-LIGHT PORCH SWITCH

A circuit for a simple night-light porch switch is shown in Figure 8.47. The idea behind the design is to be able to control lighting, indoors or outdoors, so that if the ambient light falls below a preset level, set by the variable resistor, the light will be switched on. The circuit includes a small, ac power supply that will power the light-sensing switch. An LDR (light-dependent resistor) is connected to the base of a switching transistor. If the light level is high, the resistance of the LDR is minimal, so the transistor base connection does not have sufficient voltage to turn on the transistor. When the light level falls, the resistance of the LDR increases, which in turn increases the voltage on the base of the transistor and thereby turns it on. An electrolytic capacitor is connected between ground and the base of the transistor. This

188 APPLIED SECURITY DEVICES AND CIRCUITS

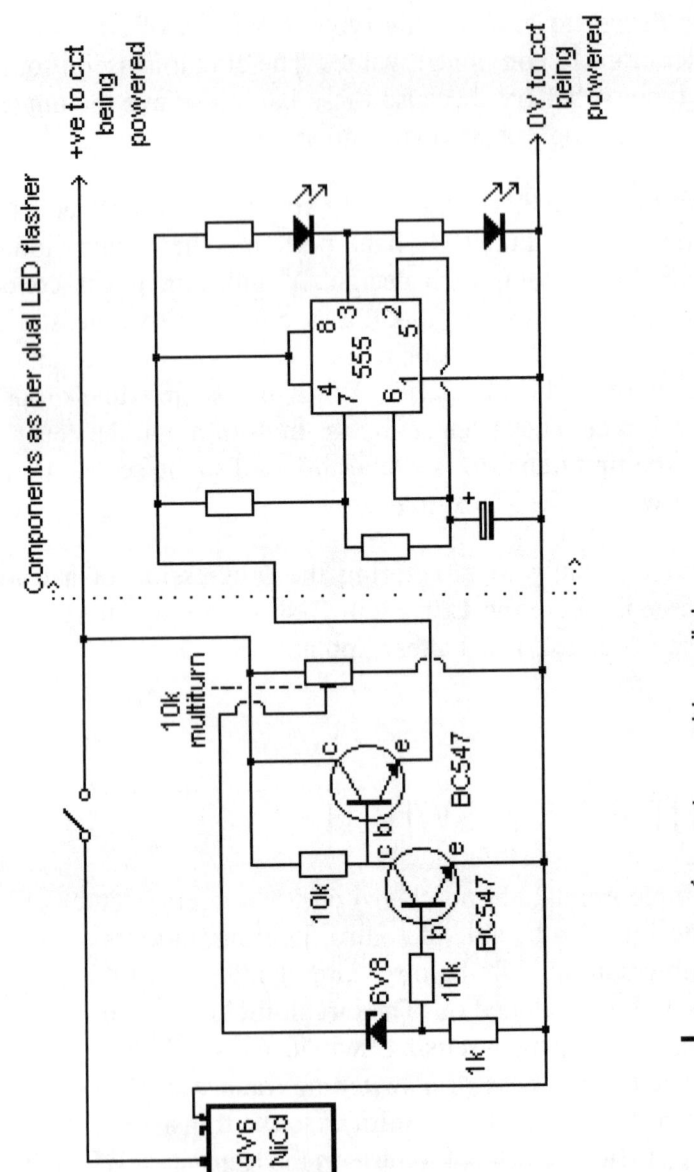

FIGURE 8.46

is because if the light level is on the edge of turning the transistor on or off, or if the voltage from the power supply fluctuates slightly as the relay turns on or off, the switching action will not shudder or vibrate intensely—often the cause of some commercial devices' continuously flashing on and off during twilight borderline light levels.

It is imperative that the power supply be switched off properly if you are changing the lamp, because if the relay switches on as a shadow is cast over the sensor, then the power supply is connected to the lamp fixture.

To prevent feedback where the illuminating lamp is pointed toward the sensor element, make sure that the sensor is sunk into the enclosure, hooded from direct light from the lamp. Another feedback problem can occur if reflective surfaces, such as a concrete wall, bounce the light from the lamp back into the sensor element, causing the lamp to switch on and off continuously during darkness.

Simple Nightlight Porch Switch

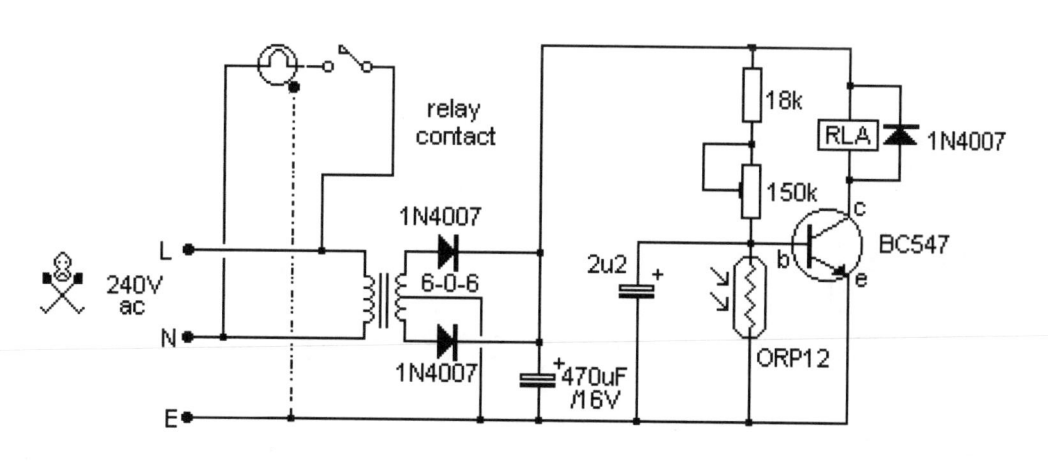

FIGURE 8.47

ONE-SECOND BEEPER/FLASHER

Figure 8.48 shows a circuit that generates a flash-rate cycle of one second on, one second off. All component pinouts are shown, because this circuit is one that anyone can enjoy constructing. It has many uses and, with only a low-component count, may be constructed in a short period of time. To give a one second on, one second off cycle, the timing capacitor has a value of 100nF. Although an output driver transistor is shown, for low-current devices to be driven, such as a single LED, the transistor perhaps may be disposed of and the LED driven directly by the output gate. Supply voltage is not critical but depends upon the type of buzzer used, which, incidentally, also may be omitted if no audible output is required.

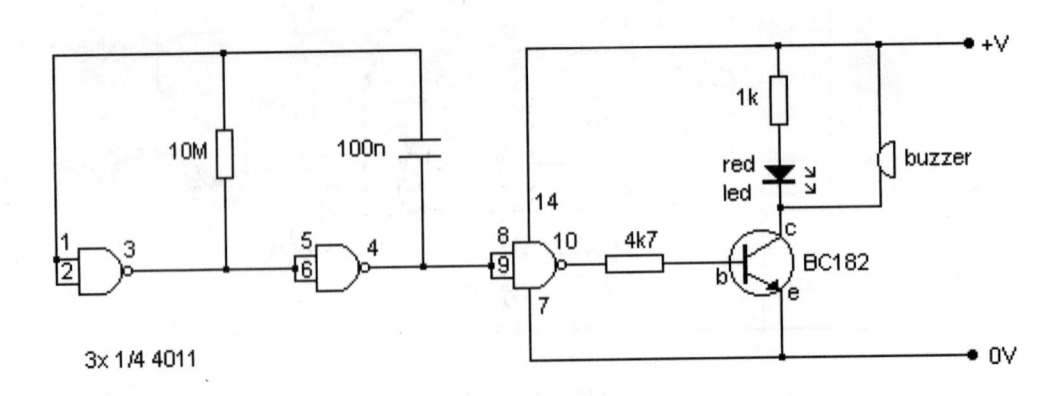

FIGURE 8.48

MISCELLANEOUS DESIGNS

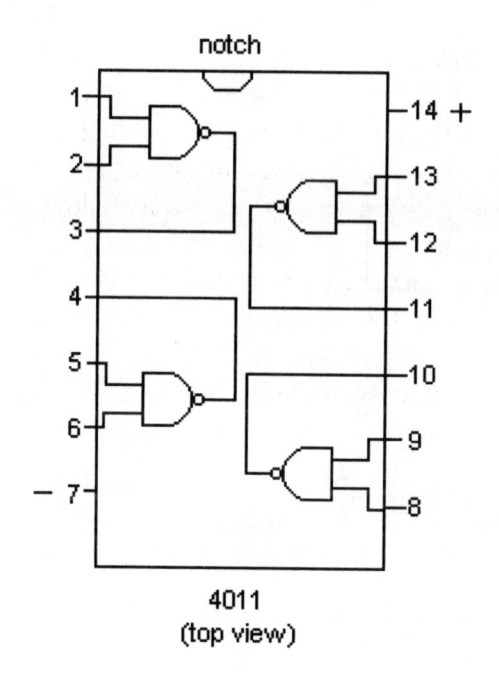

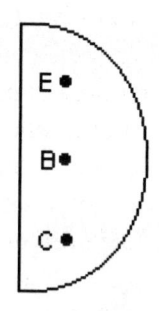

notch

4011
(top view)

pin (underside) view
of BC182

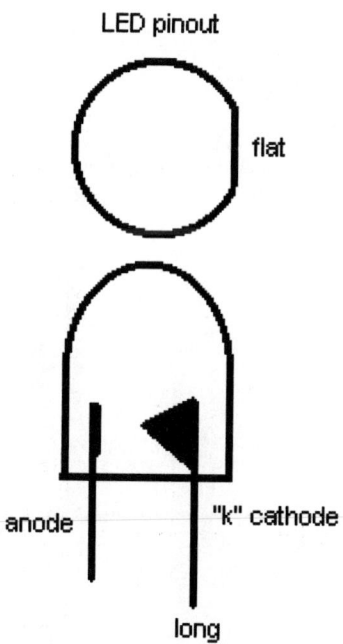

LED pinout

flat

anode "k" cathode

long

FIGURE 8.48 (Continued)

APPLIED SECURITY DEVICES AND CIRCUITS

Radio Controlled Activator

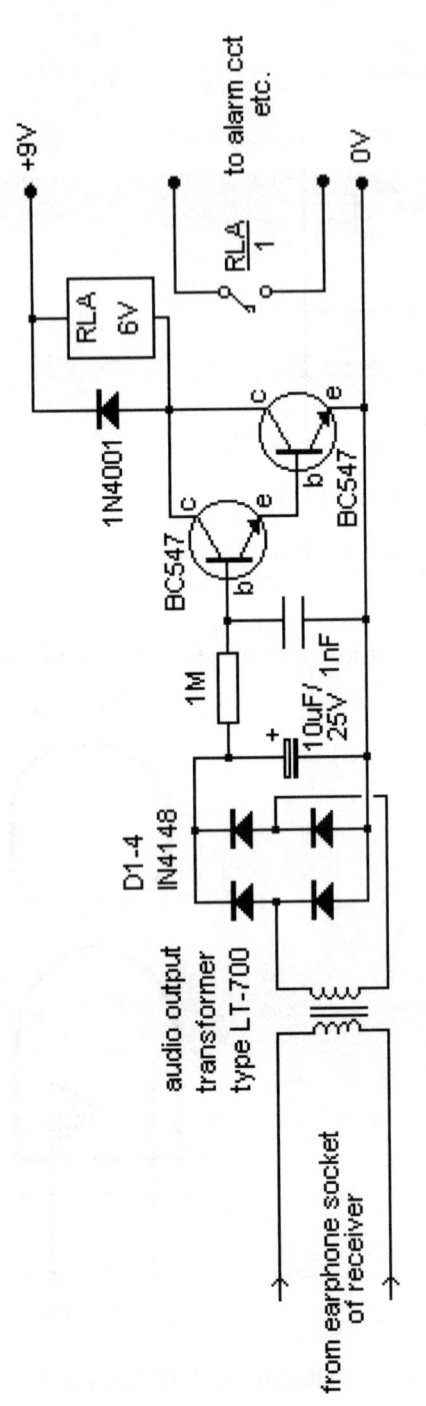

FIGURE 8.49

RADIO-CONTROLLED ACTIVATOR

A simple radio-controlled activator is shown in Figure 8.49. The idea of the design is to be able to control any device, switching it on or off, with a signal derived from the earphone socket of a radio receiver. Although there is no tone dependency, the circuit is ideal for low-budget arrangements. The transmitter can be one described in an earlier sectioncalled Discrete Monitoring, or it could be a walkie-talkie, secret telephone transmitter, or the like. The activator could control warning lights, illumination as you approach home, tape recorders, and so on.

The signal from the earphone socket is connected to a back-to-front audio output transformer, with the low-impedance side connected to the socket via a suitable plug. The signal from the transformer then is full-wave rectified and smoothed before being sent via a 1M resistor to a Darlington Pair, which then switches on the relay.

Note that the circuit requires a relatively beefy audio signal to make it actually operate, with the advantage that if the receiver does not have a squelch facility, any white noise or hash will not cause a false triggering problem.

REFRIGERATOR-LIGHT BEEPING ALARM

A design that when illuminated produces an intermittent beep is shown in Figure 8.50. The design is based upon a 14-stage ripple counter/oscillator device, with the 4060 frequency of the on/off cycle controlled by the CR combination connected to pin 9 and pin 10. The circuit operates only when the LDR is illuminated, so the resistance between pin 12 and ground falls from high resistance to just a few kilohms, depending on the LDR type used. This causes the IC to operate and start going through the cycle repeatedly until the LDR is back in darkness.

The device is contained within a plastic case with an aperture for the LDR, which is placed in close proximity to the interior switched light of the refrigerator, although lighting in the room may be sufficiently bright to trigger the

Refrigerator / Light Bleeping Alarm

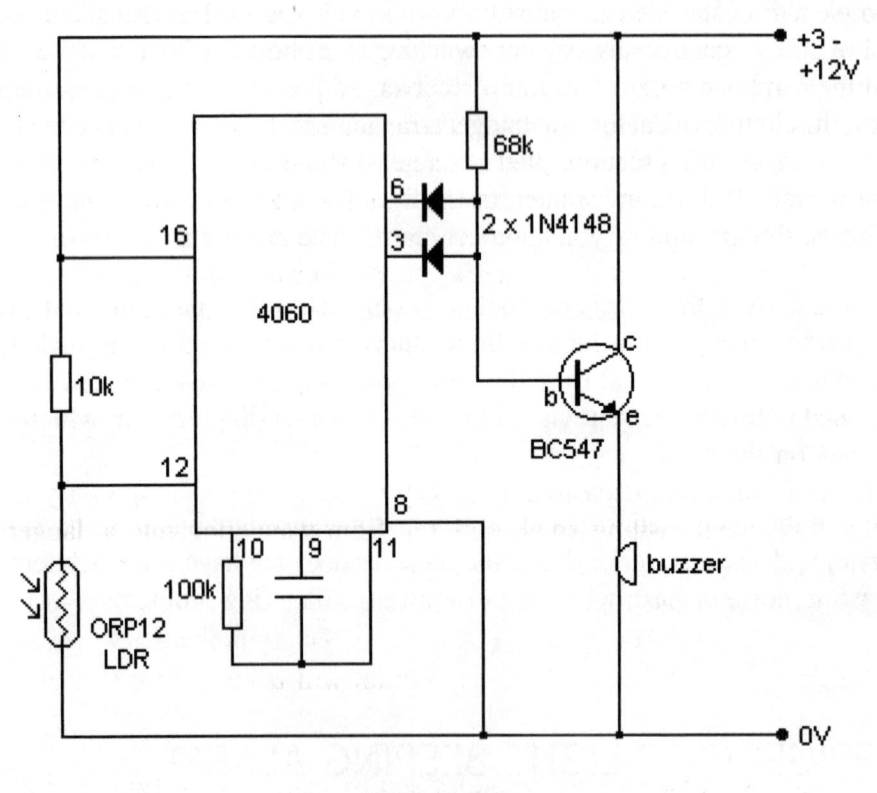

FIGURE 8.50

alarm, allowing the unit simply to be screwed to the opening door of the refrigerator.

If the whole unit is sealed in cling-style polyethylene film, moisture intrusion will be kept to a minimum; otherwise, corrosion will damage the device over time. Potting the whole circuit including a large-capacity battery is an option, with the obvious exception of a small hole for the sounder. A clear plastic lens already protects the LDR. Because current drain is minimal (depending on frequency of use), the unit could be considered a disposable device.

SNOOPER SCARERS

Snooper scarers are extremely useful in the workplace or home if you need to ward off the occasional "nosy parker" type of person, without resorting to a full-blown intruder alarm system. This type of person is the sort who might hunt around for information in an office during lunch breaks or after business hours, looking through loose paperwork in in-trays or on desks and through drawers or file cabinets for something to steal—such as your own personal stapler—or attempting to gain secrets to be used in the day-to-day political struggles that occur in every establishment. Because it is not possible to always alarm an office area due to cleaning or deliveries, or perhaps because access through the office is required at all times, this is the time when the part-time "investigator" strikes, or he or she may hang back during coffee breaks to do his or her "thing."

In the home situation, occasionally it may be necessary to temporarily alarm a room to prevent visiting small children from wandering into a dangerous situation such as a pool area or a medicine drawer in a bedside table. Maybe you occasionally suspect that a guest will pry into certain areas of your home where he or she has no business. Any area that cannot, for one reason or another, be locked or secured in any fashion will benefit from one of these devices.

The preceding problems may be addressed by using one of two methods of attack. The first is to have an alarm that gives a loud, ear-splitting output if the input device is activated. This will warn off the snooper or let the snooper know that you suspect something. The second method is to have a latched alarm that is silent and will give either a local or remote indication that the input device of the alarm circuit has been triggered. This second method of protection means that the snooper will carry out his or her devious duties unaware that they have been detected.

Silent alarms often are used in alarm systems in conjunction with panic alerts. This means that the alarm is indicated at a remote or central station without the intruder's being aware that he or she may have triggered the alarm, giving security authorities time to arrive at the scene before the villain has fled the scene. A silent alarm has the added advantage of being more environmentally friendly than an audible alarm, in the case where a false alarm triggers through an otherwise quiet night.

Two designs for silent snooper detectors, both with a latching action, are shown in Figure 8.51 and Figure 8.51a. The first design has an input for normally closed detector switches, which, if activated, cause the base of the transistor to rise and therefore switch on the transistor. The relay closes and, because one pair of the contacts is parallel with the transistor, the relay still is activated even when the transistor is switched off by closing the switch. The

FIGURE 8.51

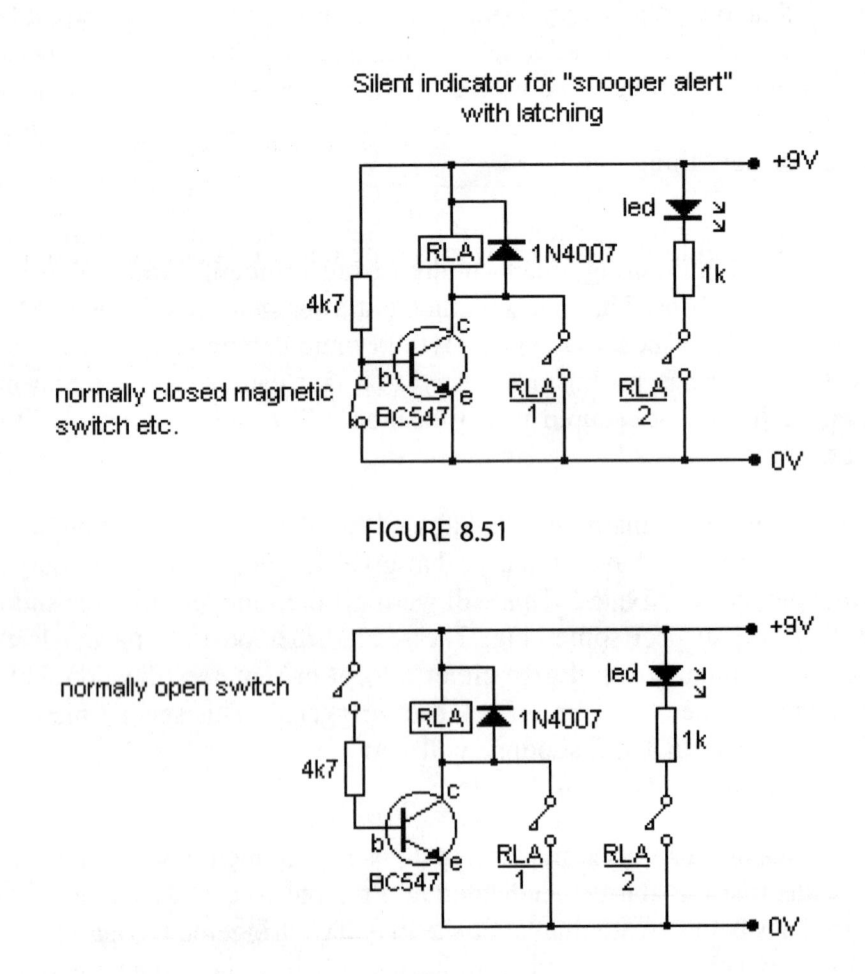

FIGURE 8.51a

other pair of contacts on the relay is connected to an LED in series with a current-limiting resistor. As with all four circuits, the relay stays latched until power is cut off, perhaps by means of a hidden key switch.

Figure 8.51a is a latching circuit that uses a system of normally open switches that, once activated, connect the base of the transistor with the supply line via a current-limiting resistor. If you find that the selected transistor has insufficient gain to switch on a particular relay, then a Darlington Pair may be used instead of a single transistor.

It is an individual decision as to which type of detector switch should be used. You may prefer fabricating one from springy pieces of brass, for example, that can be activated when papers or folders are picked up from the top of an office desk or removed from a drawer. The type of pressure switch used in musical greeting cards sometimes can be utilized. The biggest drawback to fitting snooper alarms is finding a period of time when you won't draw attention from a possible culprit! You may want a time-out circuit device that stops the alarm after a preset period.

Figure 8.52 and Figure 8.52a show designs similar to the previous ones but give an audible output to scare the intruder. If required, the audible output can

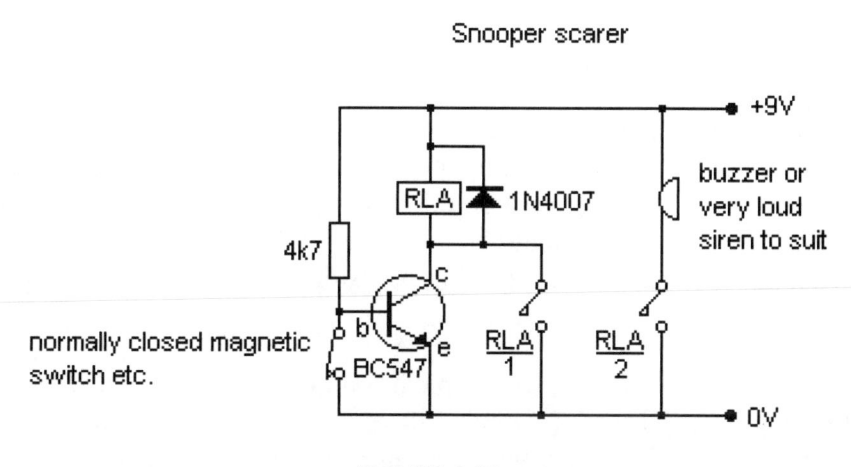

FIGURE 8.52

FIGURE 8.52a

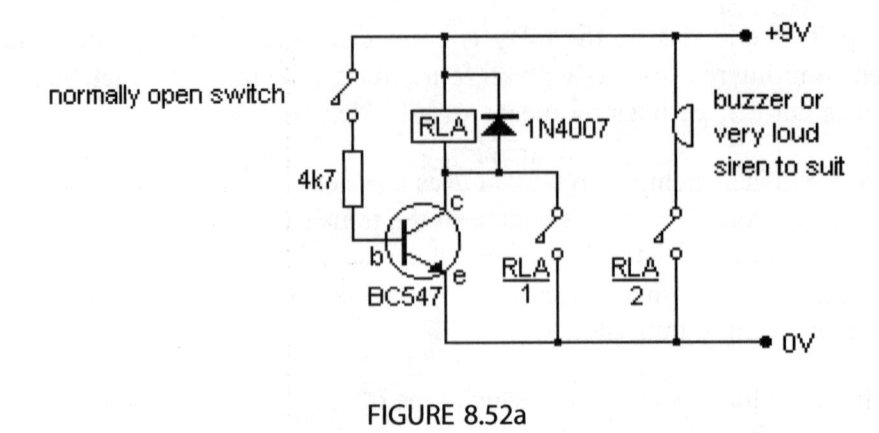

be replaced by a strobe light—or you may use both to cause a great deal of panic to the offender! If a large current is drawn by the output devices, then the supply unit should have ample capabilities, as should the switching relay contacts.

SOUND-ACTIVATED SWITCH

A sound-activated switch has many uses in security, for example, in listening for attacks to walls, floors, and ceilings in a bank prone to such attacks. Similarly, these switches may be used in cases where barking of guard dogs, who may sense potential attackers before they enter the perimeter, could trigger lights and scare off the burglars before they can cause damage. In the home, the noise of breaking glass could switch on lights or trigger other alarm indicators. The device also could be used as a VOX for tape recorders, video recorders, and any other unit requiring occasional switching that is triggered by sound.

Figure 8.53 shows the circuit for a typical sound-activated switch. The circuit has three audio-amplification circuits; these provide an ample switching signal to trigger the 555 timer, which in turn activates the relay. The preset

MISCELLANEOUS DESIGNS   199

Sound activated switch

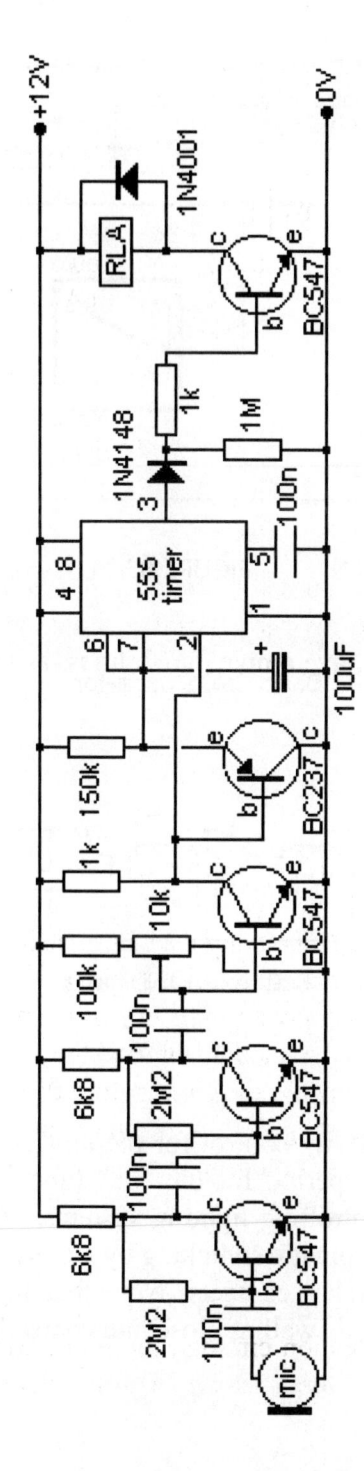

FIGURE 8.53

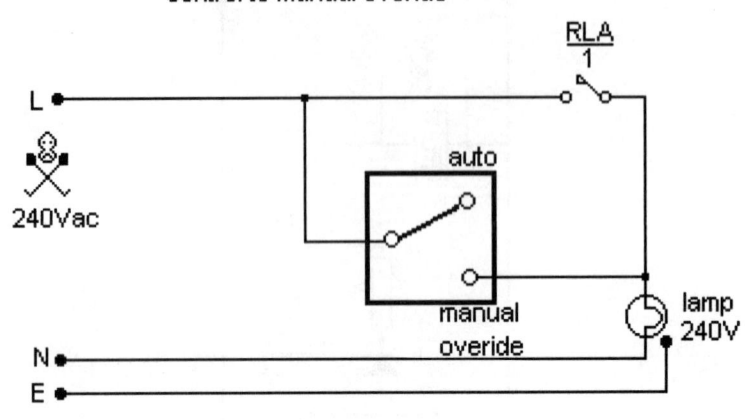

Method of switching from relay
control to manual overide

FIGURE 8.53a

Basic CMOS oscillator

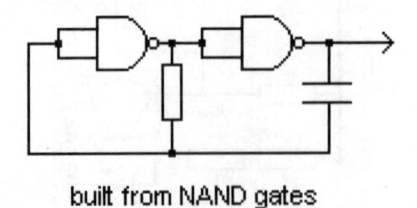

built from NAND gates

FIGURE 8.54

resistor acts as a sensitivity control for the unit, with the 100uF electrolytic capacitor introducing a period for the "on" time of the timer. Figure 8.53a shows a method of controlling lighting systems, with a way to switch from manual switching to automatic switching by simply wiring a switch across the n.o. (normally open) relay contacts. Note that for safety reasons, another power isolation switch, as well as close and coarse fuse protection, should be employed.

High powered pulsing tone

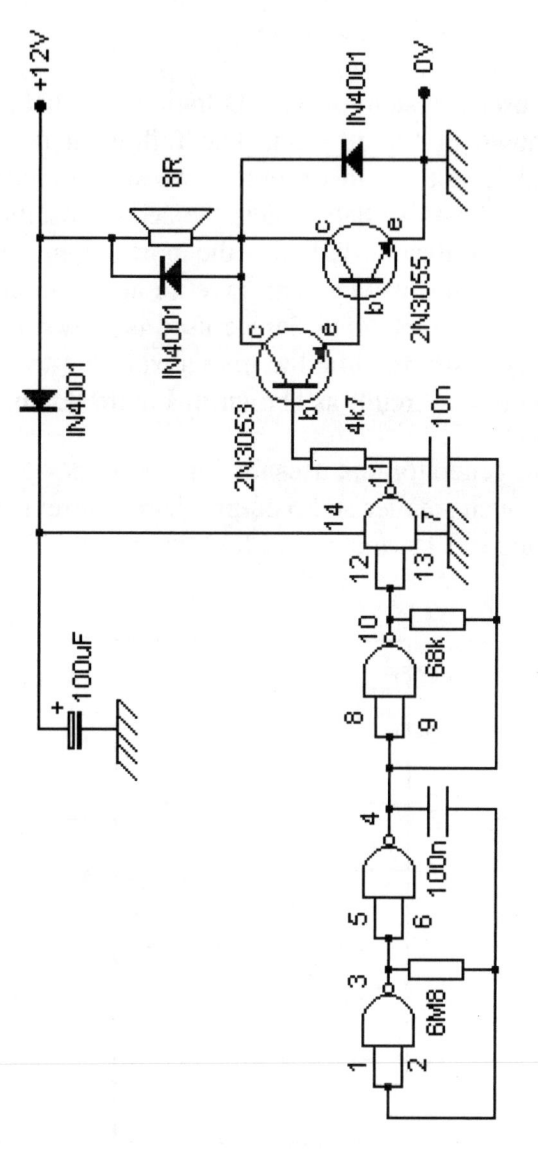

FIGURE 8.55

SOUNDER CIRCUITS

This section introduces various designs suitable for driving loudspeakers as audible warning devices in any system.

The first circuit uses a system of NAND logic gates that are connected as an oscillator, as shown in Figure 8.54. The following diagram, Figure 8.55, shows a practical design that incorporates two separate oscillators that give a pulsing tone as the first oscillator gates the second oscillator. The output of the second oscillator then is fed into the power amplifier transistors and, depending on the supply voltage, can give an audio output of several watts. The top view of the 4011 CMOS device used is shown in Figure 8.55a. The circuit can be very easily modified to give a warbling-tone output by making alterations to the basic circuit, as shown in Figure 8.55b.

A two-tone audio generator that uses the TTL 7413 device is shown in Figure 8.56. This has a much smaller audio output than the previous two circuits, and because the device is TTL and not CMOS, the supply voltage to the IC must

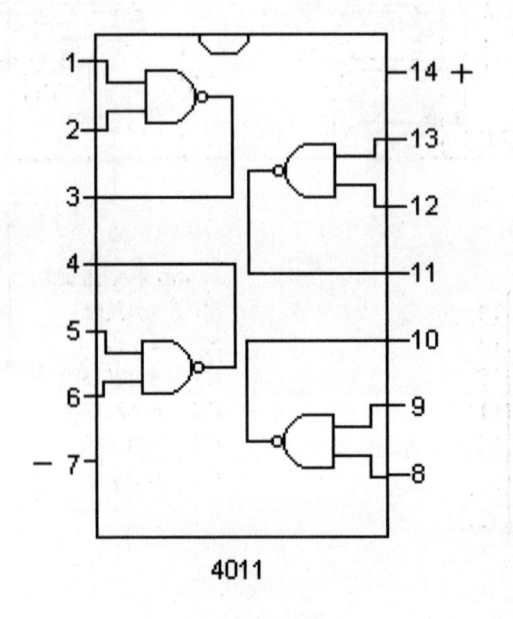

FIGURE 8.55a

MISCELLANEOUS DESIGNS 203

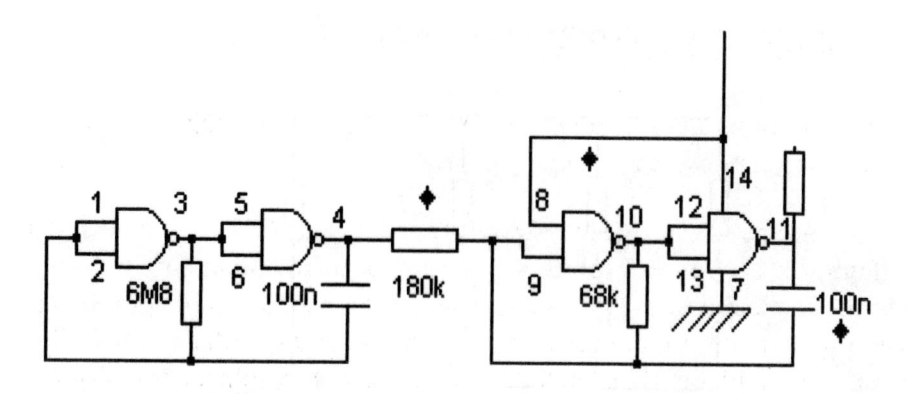

FIGURE 8.55b "Warble" effect modification

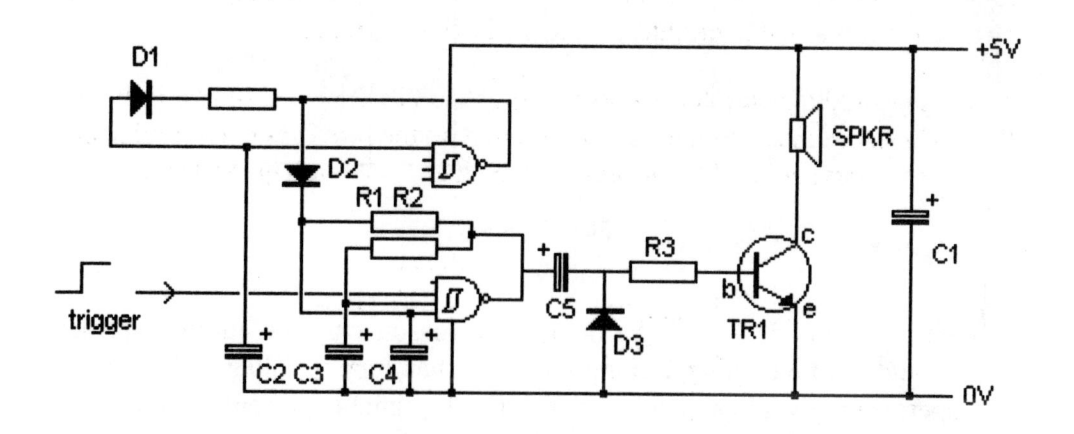

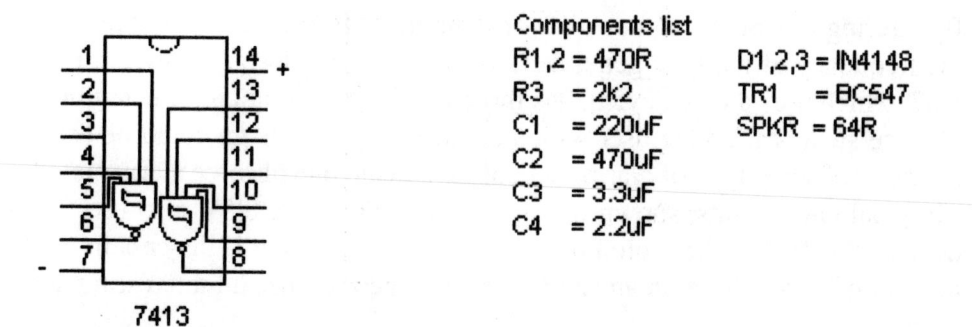

Components list

R1,2 = 470R
R3 = 2k2
C1 = 220uF
C2 = 470uF
C3 = 3.3uF
C4 = 2.2uF

D1,2,3 = IN4148
TR1 = BC547
SPKR = 64R

7413

FIGURE 8.56 Two-tone audio generator

Three-tone signal using a UJT

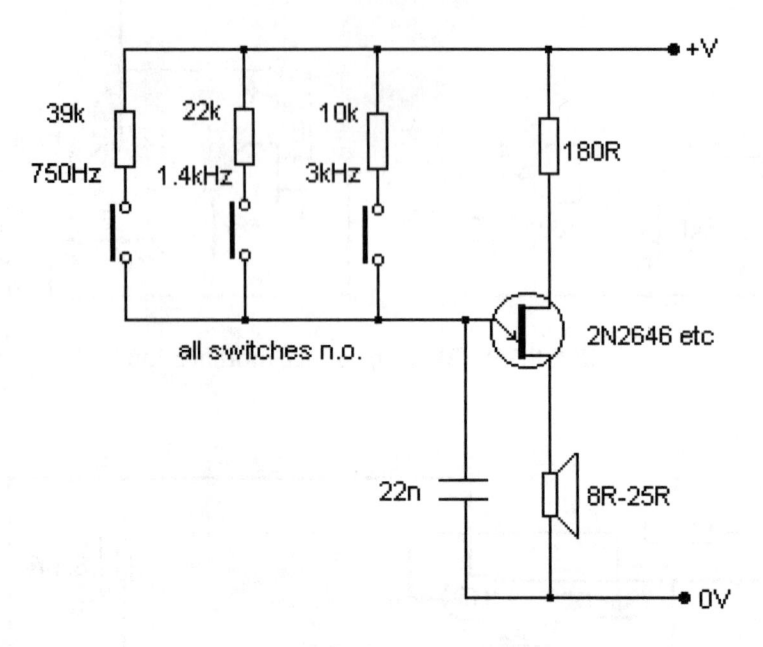

FIGURE 8.57

be kept to around 5V, although there is no reason why the audio amplifier stage could not be upgraded along with a higher supply voltage. The circuit will operate only while a positive-going trigger signal is present, such as may be obtained from an existing alarm system, as indicated.

By altering one of the values of C or R in an oscillator, it is possible to alter the frequency of the tone generated. To this end, the design shown in Figure 8.57, which uses a UJT device, has three tone outputs that are selected by one of three switches. To the user with a musical ear, the design could be used as the basis of some type of alarm that could indicate one of three alarm statuses. The final two circuits, shown in Figure 8.58 and Figure 8.59, also use a UJT device. The first of these circuits has an audio output that is dependent on the amount of light falling on an LDR; the second circuit has an interesting audio output!

MISCELLANEOUS DESIGNS | 205

Light-sensitive oscillator using a UJT

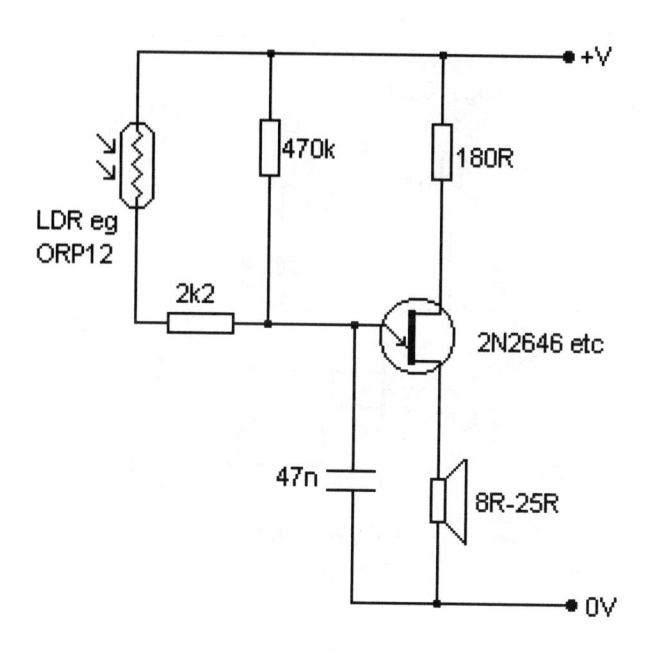

FIGURE 8.58

Rising frequency oscillator using a UJT

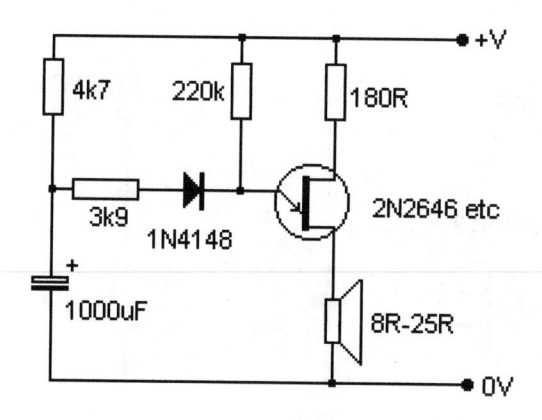

FIGURE 8.59

Simple Static Detector

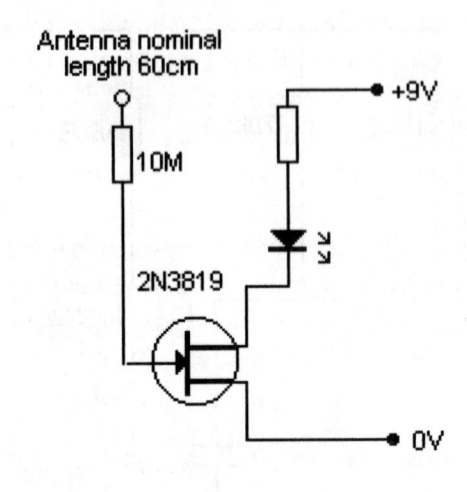

FIGURE 8.60

Static electricity / lightning detector

FIGURE 8.60a

STATIC ELECTRICITY/LIGHTNING DETECTORS

Two simple static electricity detectors are shown in Figures 8.60 and 8.60a. Both have an LED output that will illuminate when a sufficiently large field of static charge is sensed by a short pickup antenna. The second circuit also can be used as a sensitive lightning detector. The first circuit is constructed with an aerial length of around 30cm. If the module is approached slowly, with a hand edging toward it, the LED will illuminate if a sufficient static charge is on the body. If the shoes of the person are wiped on a nylon carpet, the LED will go on and off with each wipe. The device can be used to show people who handle electronic components in a manufacturing environment the static charge they are carrying or generating, thereby helping them visualize the importance of wearing suitable antistatic clothing and equipment.

If either of these devices is used as a detector for long-distance lightning discharges, it is not safe or necessary to erect high aerials that may become lightning conductors. If a large aerial is available, such as 20m or so of insulated hookup wire wrapped around the loft space, the simplest lightning detector—albeit not as sensitive as the first two examples—is shown in Figure 8.60b. This includes a neon bulb, such as that contained inside a power supply voltage screwdriver/tester, and a sensitivity control if required. If it is not desirable to pull a neon screwdriver apart, then simply use the screwdriver, as shown in Figure 8.60c, when a storm is very close, in a darkened room!

TELEPHONE RINGER EXTENDER

Very often, it is impossible to hear a telephone ringing for one of the following reasons:

1. The person is hard of hearing.

2. Ambient noise in a workshop or factory is so high that it overrides the ringing sound.

208 APPLIED SECURITY DEVICES AND CIRCUITS

Short Range Lightning Detectors

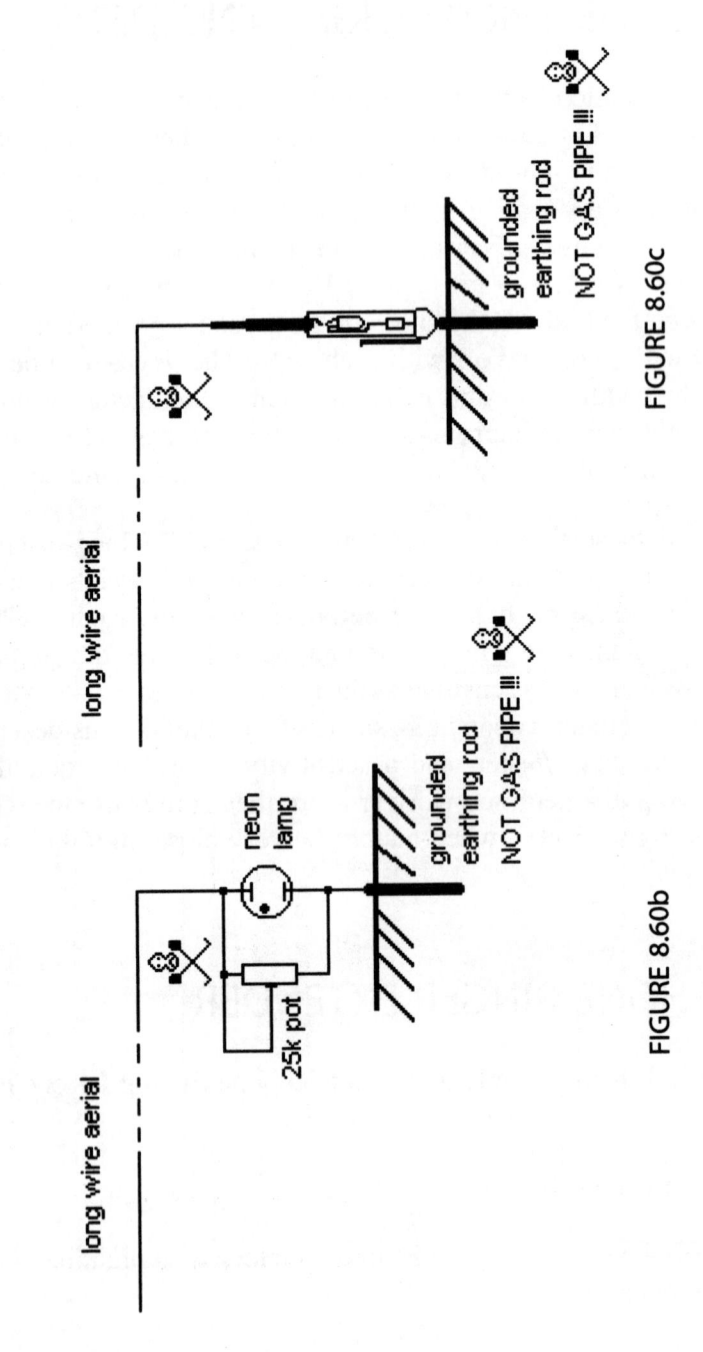

long wire aerial

neon lamp

25k pot

grounded earthing rod
NOT GAS PIPE !!!

FIGURE 8.60b

long wire aerial

grounded earthing rod
NOT GAS PIPE !!!

FIGURE 8.60c

MISCELLANEOUS DESIGNS 209

Telephone Ringer Extender

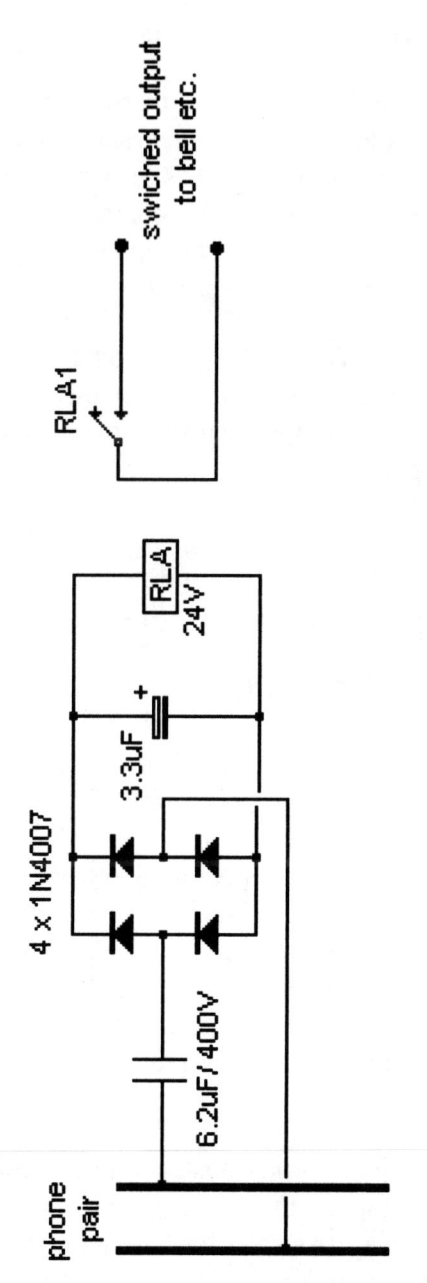

FIGURE 8.61

3. The telephone is too far away for the person to hear it, perhaps in a dwelling where the person is in a remote garage, for example.

The design in Figure 8.61 shows an interface that may be connected between the telephone line and a remote ringer, such as another bell or sounder. The relay will chatter at the same speed as the ringer, therefore some kind of extra circuit may be connected to the relay contacts to give a smooth on/off switching signal if required. In addition, a lamp circuit may be connected to the relay switch if a visual indicator is required.

Note that in many areas it is illegal to connect any nonapproved circuit to a telephone line; therefore, if you connect this design to a private telephone exchange system without approval from the appropriate authorities, you well may be breaching the law.

TEMPERATURE-ACTIVATED APPLIANCE SWITCH

The circuit for a typical temperature-activated switch is illustrated in Figure 8.62. It is very similar to the porch-lighting device shown in Figure 8.47 in the earlier section Night-Light Porch Switch, with the following exceptions. The sensing element is a negative temperature coefficient thermistor that is placed, along with the temperature set potentiometer, at a suitable location. The thermistor is connected to the noninverting input of a 351 op-amp device so that, if the resistance of the thermistor increases, the op-amp switches on the relay driver transistor. For safety reasons, both the neutral and live lines to the appliance socket are switched by independent relay contacts.

If you use this device as a thermostatically controlled heating switch, bear in mind that positioning of the sensing element is important, because if the sensor is too far away from the heat source, or heated area, then the sensor always will be in the cold and so continuously thinking that the switch needs to be on.

MISCELLANEOUS DESIGNS 211

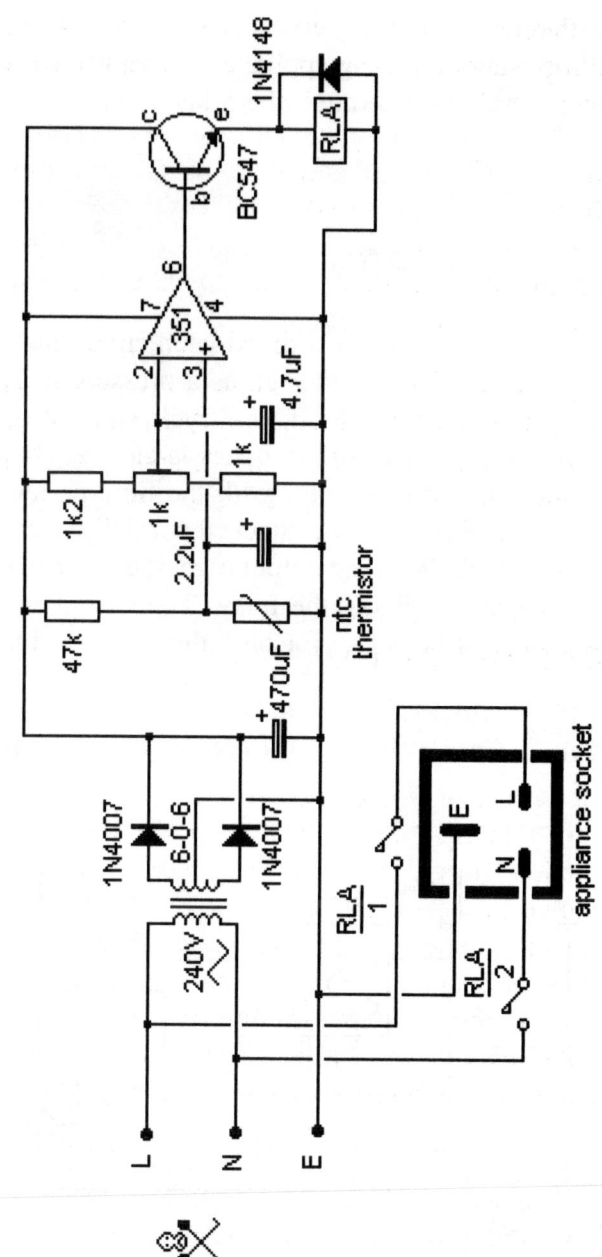

Temperature activated appliance switch

FIGURE 8.62

The relay, with double-switching contacts, must be of sufficient contact rating because, by implication, heating appliances generally are current-hungry and so may draw several amps through the relay contacts.

TIMED LATCH ALARM

The circuit in Figure 8.63 shows a timed latch circuit that is operated by one or more normally open switches, such as a pressure mat, for example. The design is based on a standard 555 IC one-cycle circuit that is triggered at pin 2 by being sent low if any of the switches is closed. The output of the 555, pin 3, stays high until it times out, with the time period depending on the values of both C and R, even if the attacked normally open switch is reopened straight after the attack. With the output high, the npn driver transistor will be turned on, therefore energizing the relay. The design is not a true latching circuit, because it will be active for only the timed period.

Timed latch alarm operated by a normally open switch

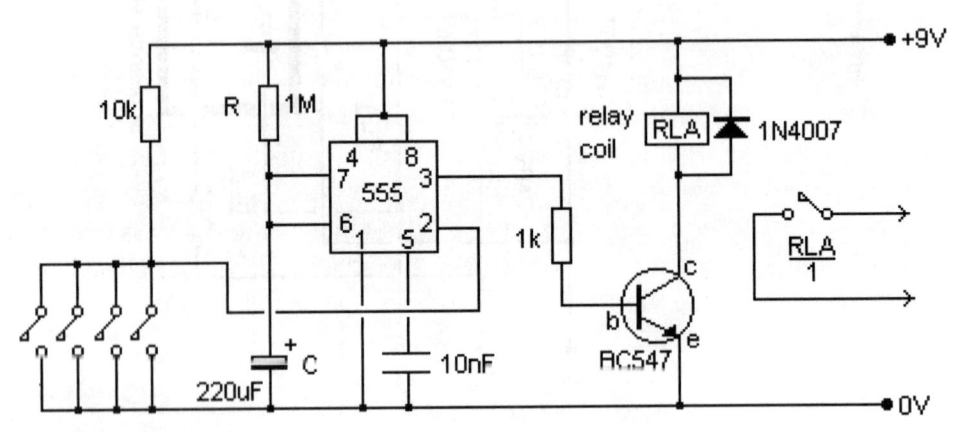

paralled normally open switches eg. panic switch pressure mat etc.

FIGURE 8.63

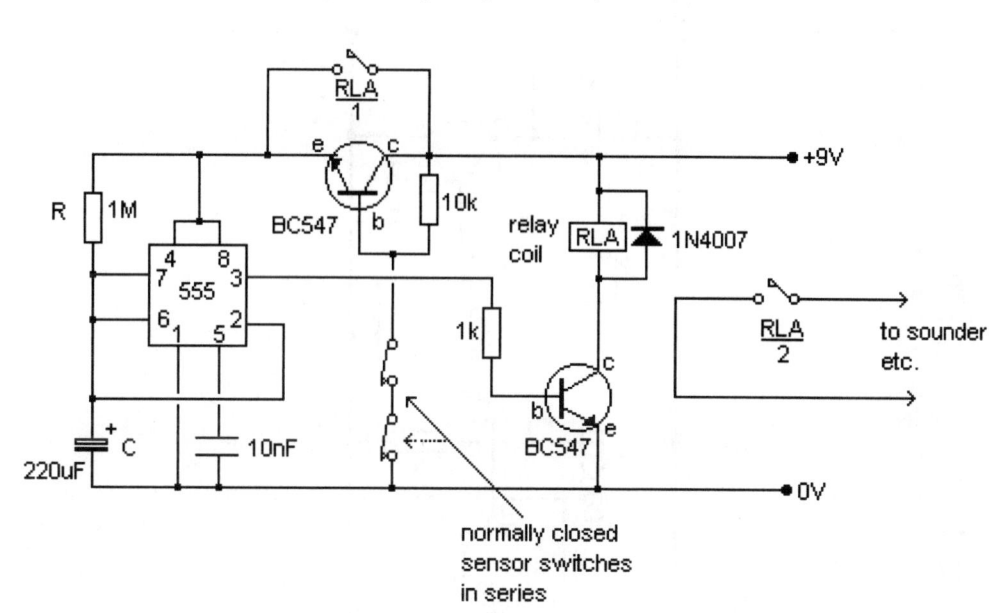

Timed latch alarm operated by multiple normally closed switches

FIGURE 8.64

Figure 8.64 illustrates a timed latch circuit that is operated by one or more normally closed switches. Power to the timing circuit happens only if one of the switches is attacked and becomes opened. The transistor in series with the supply line then switches power on to the timer, whose output stays high for a time period depending on the value of C and R. Because the power-switching transistor is shorted out by a pair of relay contacts across the emitter and collector for the timed period, even if the attacked normally closed switch is closed quickly, the alarm still is active for the timed period.

TOUCH-ACTIVATED ALARM

A design for a touch-activated alarm is shown in Figure 8.65. A problem often encountered in a touch-switch design is false triggering, so I recommend that

214 APPLIED SECURITY DEVICES AND CIRCUITS

FIGURE 8.65 Touch activated alarm

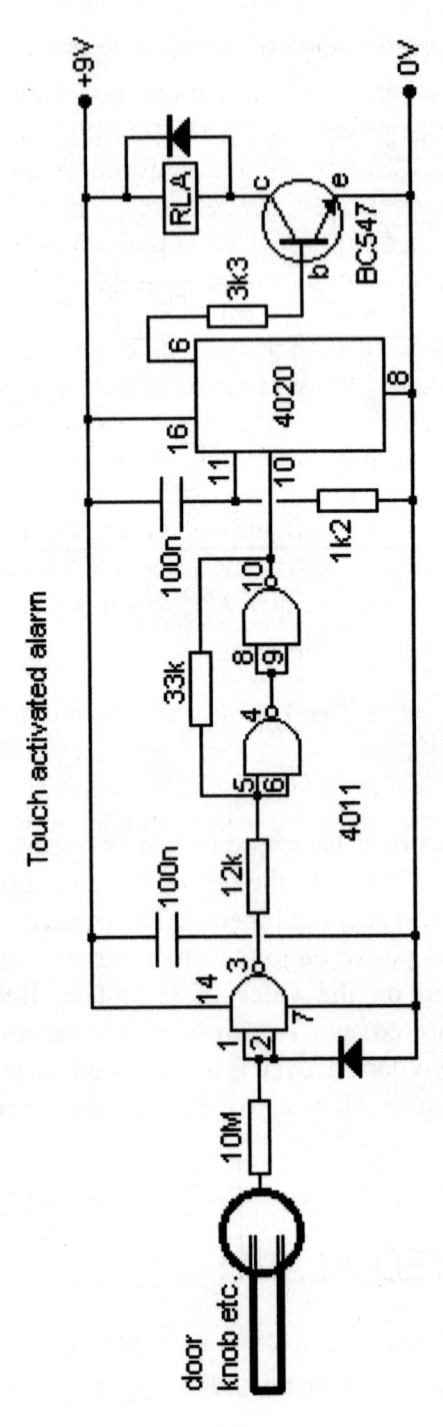

some kind of pulse-counting be utilized. Taking this approach means that the circuit will not give an output until a certain number of pulses is detected. To this end, a 14-stage ripple counter, the 4020 CMOS IC, is used. On powering up the circuit, circuit reset on power-up is achieved by the fact that the reset pin, pin 11, is virtually connected to the positive supply line by a 100nF capacitor. When a sufficient number of input pulses has been reached on pin 10, the selected output pin then goes high, turning on the output transistor.

To provide a latching action to the alarm, you can use a relay as shown in Figure 8.65a. This will mean that, once the relay is energized, two of the relay contacts close, keeping the relay energized. The remaining contacts of the relay are used to operate an indicator or alarm circuit.

Because it will be necessary to switch off and on the power to the entire circuit in order to obtain a reset, you will want a method of silencing the alarm. Figure 8.65b shows how to achieve this by inserting a push-to-open switch in series with the latching relay contacts.

The design works on the principle of a person's touching the input wire and injecting power supply hum into the circuit; therefore, the effectiveness of the device depends on the actual presence of ac-powered wiring in the vicinity of the alarm! The input wiring from the sensor should be as short as possible,

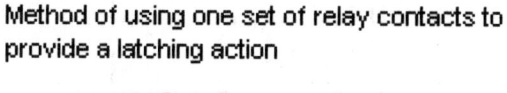

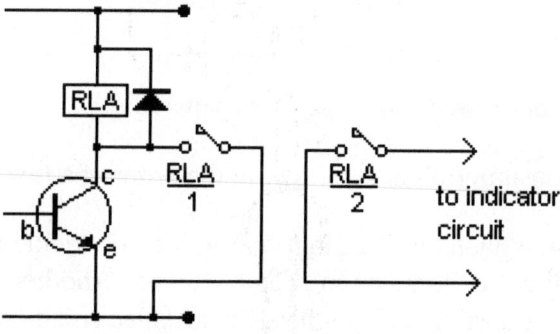

FIGURE 8.65a

Method of using a normally closed switch to act as a reset button

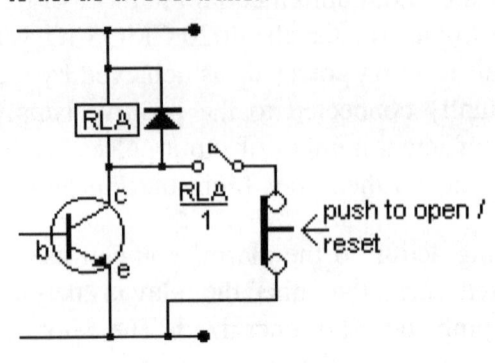

FIGURE 8.65b

isolated from ground, and with any wiring from the door handle, for example, preferably constructed from screened cabling.

TRANSFORMERLESS POWER SUPPLY

The idea of using transformerless power supplies has been around quite some time. They lend themselves to any particular application where the following are true:

▲ A power supply is required.

▲ Transformers prove too large or expensive.

▲ Current consumption is very low, in the order of a few milliamps or so.

Typical applications include charging up batteries or trickle-charging backup batteries or low-power alarm or indicator circuits. Another use is powering secret surveillance transmitters, which can be installed and then left indefinitely, without having to return to the scene in order to replace worn-out batteries. The

Transformerless low current mains supply PSU

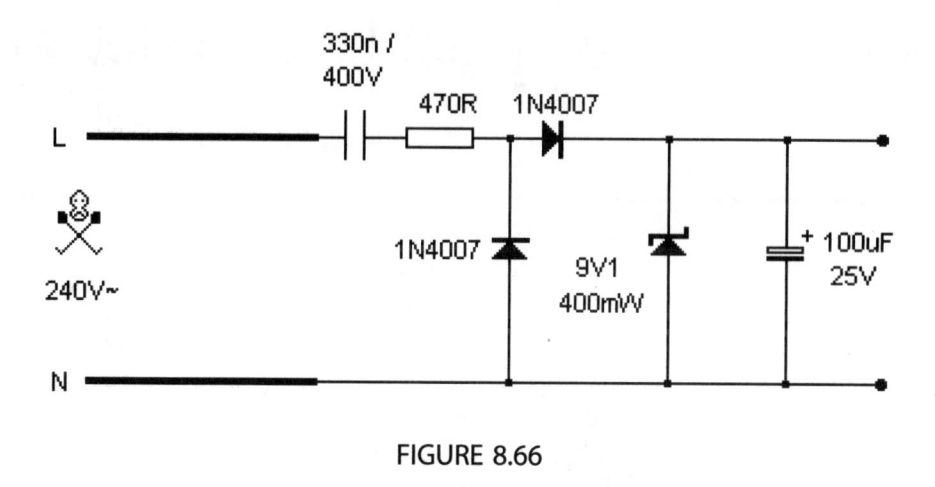

FIGURE 8.66

idea of the transformerless supply, generally speaking, is based on the ability of a capacitor to seem to pass an AC, albeit reluctantly, due to the device's inherent capacitive reactance. This ability is used to connect a high-voltage power source to a capacitor, whose value is chosen specifically to allow a small amount of current through without having the large power/heat/size problems that occur when you try to perform the same task by using power-dropping resistors.

Three slightly different circuit designs are shown in Figure 8.66, Figure 8.66a, and Figure 8.66b. Note that the capacitor used for dropping the power supply voltage definitely must have a sufficient AC working voltage, because any breakdown will cause either the capacitor to blow into an open circuit or, as is usually the case, an instantaneous short circuit, placing power supply potential onto the circuit being powered and thus causing a good deal of component damage.

You must take caution when working on these circuits, because if the live and neutral wires are mixed up, then the 0V line is at mains potential and therefore potentially lethal.

FIGURE 8.66a

FIGURE 8.66b

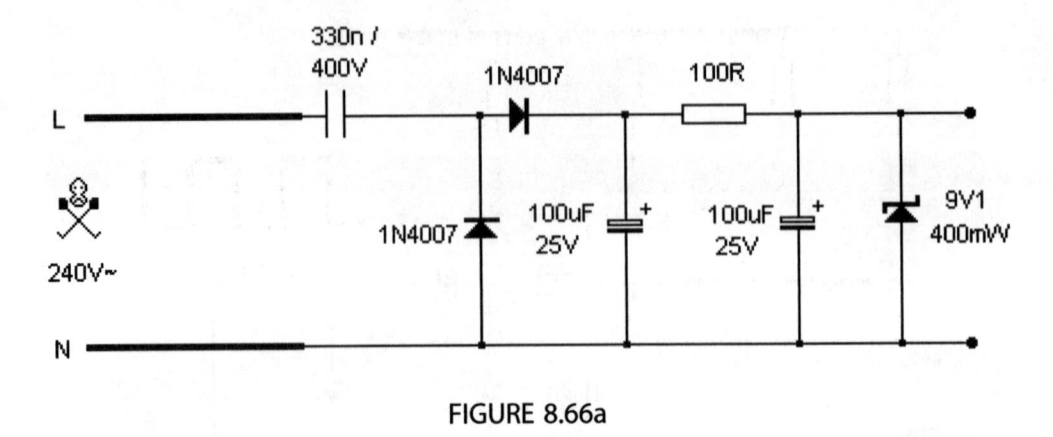

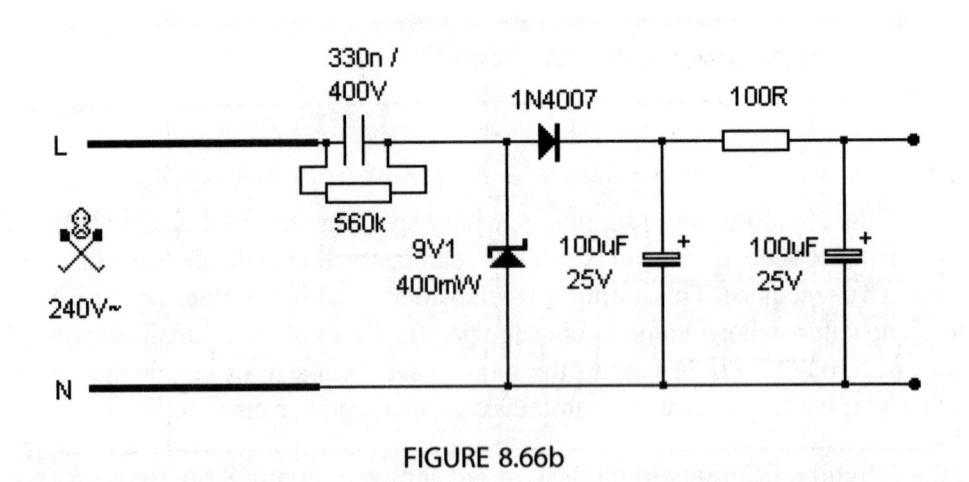

TWO-TRANSISTOR MULTIVIBRATOR CIRCUIT

As seen in Figure 8.29 in the earlier section on the Fishbite Alarm, a multivibrator circuit can be built using two npn transistors. This circuit proves useful on occasions when a 555 timer IC is not available. The frequency of operation is controlled by the two RC circuits, with some effect on the timing's being caused by excessive loading from having a small resistance in

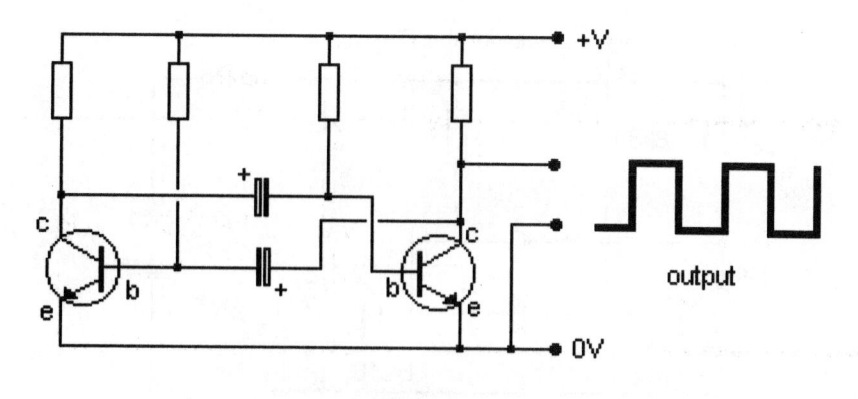

FIGURE 8.67

Multivibrator with driver/buffer

FIGURE 8.68

the collector circuits. The circuit also may have a variable mark/space ratio by altering the value of the C or R in one of the legs. If large-value capacitors will be used, then they will be electrolytic or tantalum devices—that is, polarized—so it is important to note their polarity. The output of this circuit is pretty much a square wave (see Figure 8.67) whose value will swing from nearly Vs to nearly 0V. The output from the circuit can be taken from either collector of the two transistors.

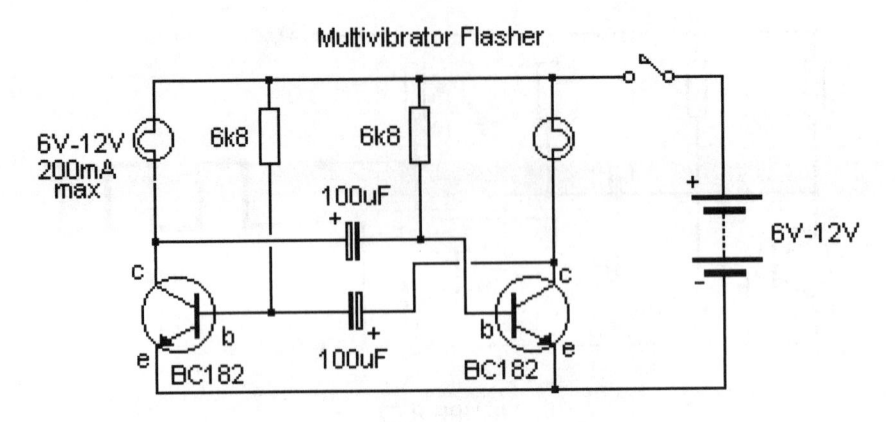

FIGURE 8.69 — Multivibrator Flasher (6V–12V, 200mA max; 6k8 resistors; 100uF capacitors; BC182 transistors; 6V–12V supply)

If a large load will be driven, such as a powerful filament lamp for an emergency beacon, a further transistor capable of taking heavy current may be used, as shown in Figure 8.68. Alternatively, this third transistor could be used to switch a relay whose contacts then would handle large loads. A design that alternately will flash on and off two filament bulbs is shown in Figure 8.69. The two transistors could be upgraded to heavier devices, but heat sinks may be required to prevent damage to the transistors.

USING LEDS AS HIGH-POWER INDICATORS

The humble, basic LED in its own right is not very visual in other than poor- or low-light levels. High-output LEDs are available on the market; however, I came across a large bag of standard devices at a HAMfest and so decided to experiment with them. Although it is possible to parallel any number of LEDs together, I carried out an experiment to see how many devices could be strung together in series, as shown in Figure 8.70. The supply voltage was a regulated 13.5V, and I found that six devices worked so long as a current-limiting resistor of 100R was included in the chain. Deciding to do away with the limiting resistor, I found that the chain worked very well with an additional LED included.

MISCELLANEOUS DESIGNS 221

Using LED's as High Power Indicators

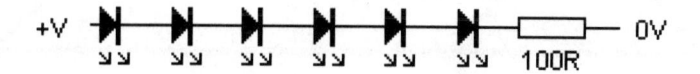

FIGURE 8.70

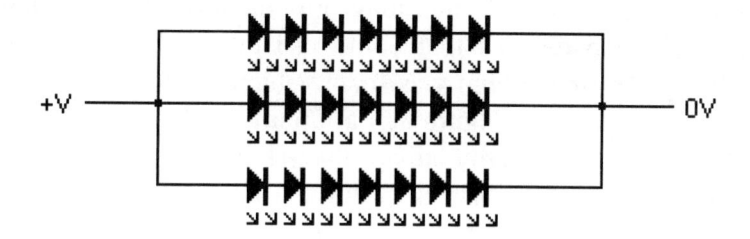

FIGURE 8.70a

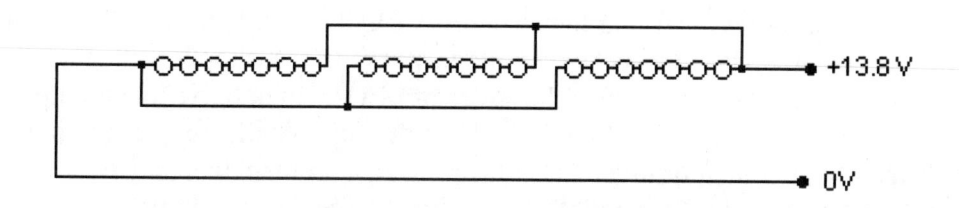

FIGURE 8.70b

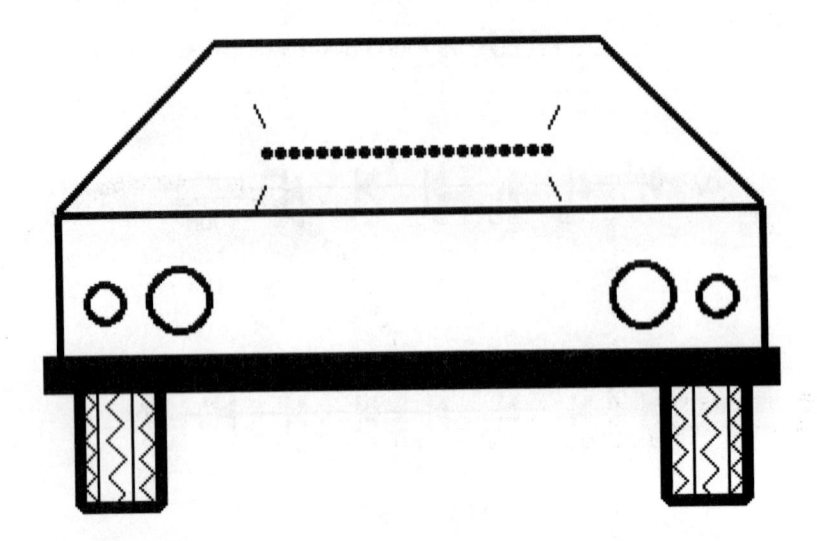

FIGURE 8.70c

If one chain of LEDs is used, this may provide sufficient brightness, but I decided to build two designs that could use these LEDs, with additional parallel chains. The first design was for a vehicle rear-windshield eye-level braking indicator, shown in Figure 8.70a, Figure 8.70b, and Figure 8.70c. In all, 21 LEDs were used, with my only problem being mounting the LEDs; this was overcome by using a few strips of thin plastic sheet, sprayed with glossy black paint to make the display radiate extremely brightly!

The second use that the surplus LEDs were put to was for an outdoor alarm box. An expensive strobe was needed to complete building the bell box, but three rows of LEDs, shielded from the elements by the built-in plastic lens of the box, were wired to the strobe output of the alarm and proved adequate.

An important point regarding the stringing of LEDs in this fashion is that of supply voltage. The idea is very voltage-wary. If the supply voltage falls below the working voltage, then the string of LEDs will not give much output, or even will not light at all. Alternatively, if the supply voltage goes too high, the smell of popping/burning LEDs may prevail, a problem that may not occur if the LEDs are wired in parallel, with either a large-wattage limiting resistor or individual resistors in series with each LED.

Infra-red LED "Spotlight"

infra-red
led cluster

automobile
headlight
reflector

infra-red
led cluster

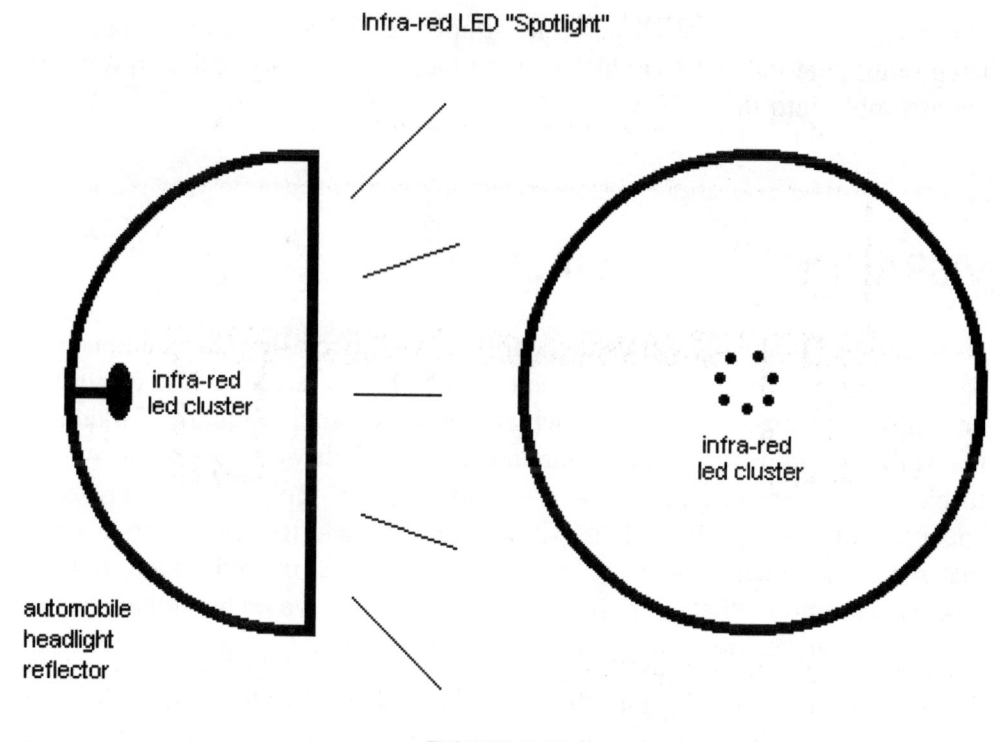

FIGURE 8.70d

INFRARED "NIGHT-VISION" ILLUMINATOR

Another use of multi-LED arrangements is shown in Figure 8.70d. This illustrates the assembly of a crude but low-cost infrared spotlight for illuminating areas that are covered by a closed-circuit CCD TV camera. Infrared LEDs often are built in to commercially available cameras so as to provide night vision, but they will light up the face only of someone standing two or three feet from the unit. With the device shown, a much larger area can be illuminated with infrared, unseen by an intruder but picked up by a CCD-style camera.

The reflector lens cone is an old automobile headlamp unit picked up from a junkyard. It is best to obtain one with a lamp bulb still in its original place, so that the infrared LED cluster can be mounted in the correct position to obtain a good beam. If at all possible, simply break the glass of the old 12V lamp bulb, glue in the LED cluster to the base of the bulb, and then reinsert the assembly into the reflector.

VIBRATION ALARM CIRCUITS

As mentioned in the section listing alarm input devices, one cheap and useful device is a vibration, tilt, or movement switch. The type used in the prototype design was a mercury tilt switch, which contains a small amount of mercury that rolls up and down a small container where, if tilted, it causes the switch to close. Mercury switches now are being rapidly replaced by more eco-friendly ball bearing types. Tilt switches happily lend themselves to designs that sense movement when the axis of the switch is moved, albeit slowly, whereas vibration—that is, trembler—switches are activated by movements in any plane or direction so long as the movement is quick enough to cause the contacts to close. The following four circuits may be used with any type of normally open input switch.

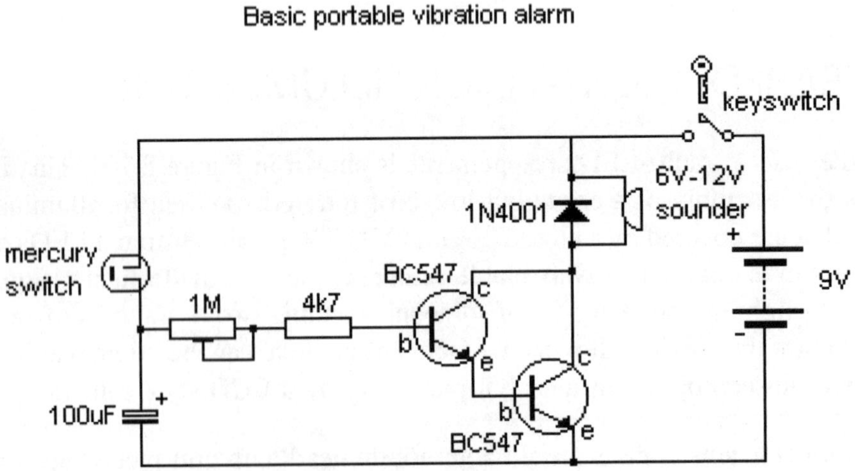

Basic portable vibration alarm

FIGURE 8.71

Figure 8.71 shows a basic portable vibration alarm that can be assembled cheaply. When the switch is closed even for a brief moment, the capacitor immediately charges up through the switch. If the capacitor is too large a value, there may be danger of causing arc damage to the switch contacts, which will lead to a malfunction. The capacitor now instantly begins to discharge through the variable resistor, which acts as a rough timing method, to the Darlington Pair, which switch on the sounder. Once the capacitor is sufficiently discharged, the circuit ceases sounding. A key switch in series with the battery will enable the unit to be armed when necessary.

Figure 8.72 shows the same circuit, this time with a relay output that gives the opportunity to drive a larger load than the transistor output design is capable of; however, bear in mind that the circuit is battery powered and so excessive current drain soon will flatten a dry-cell battery. Note that in the previous circuit, the cutoff period—that is, when the capacitor is almost discharged and starting to cut off the output transistors—is not a straight line. This means that the buzzer or sounder at this point probably will make a rather sick sound just before silencing. Adding a relay removes this problem, because the relay cutoff will be sharp.

The design may be modified again by adding a relay with extra contacts, as shown in Figure 8.73, so that instead of the alarm's sounding for only a short

Portable vibration alarm with relay output

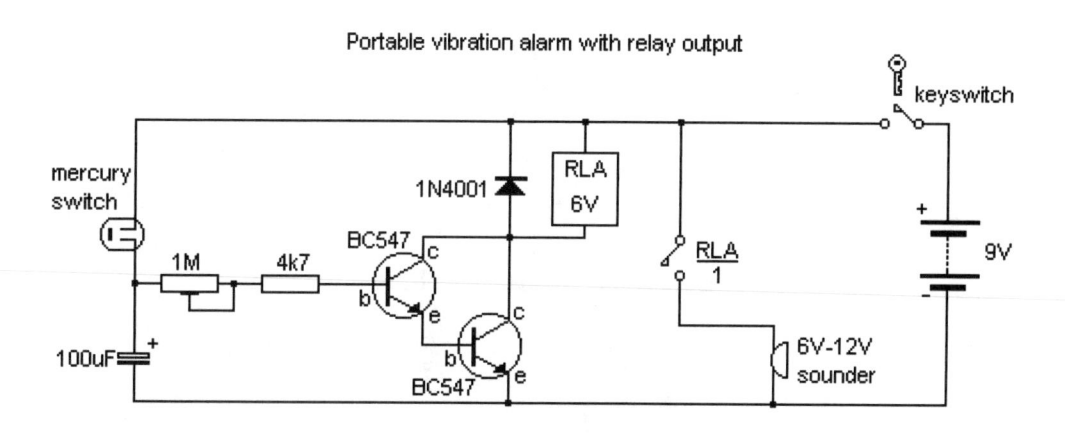

FIGURE 8.72

226 APPLIED SECURITY DEVICES AND CIRCUITS

Vibration alarm with latch and reset switch

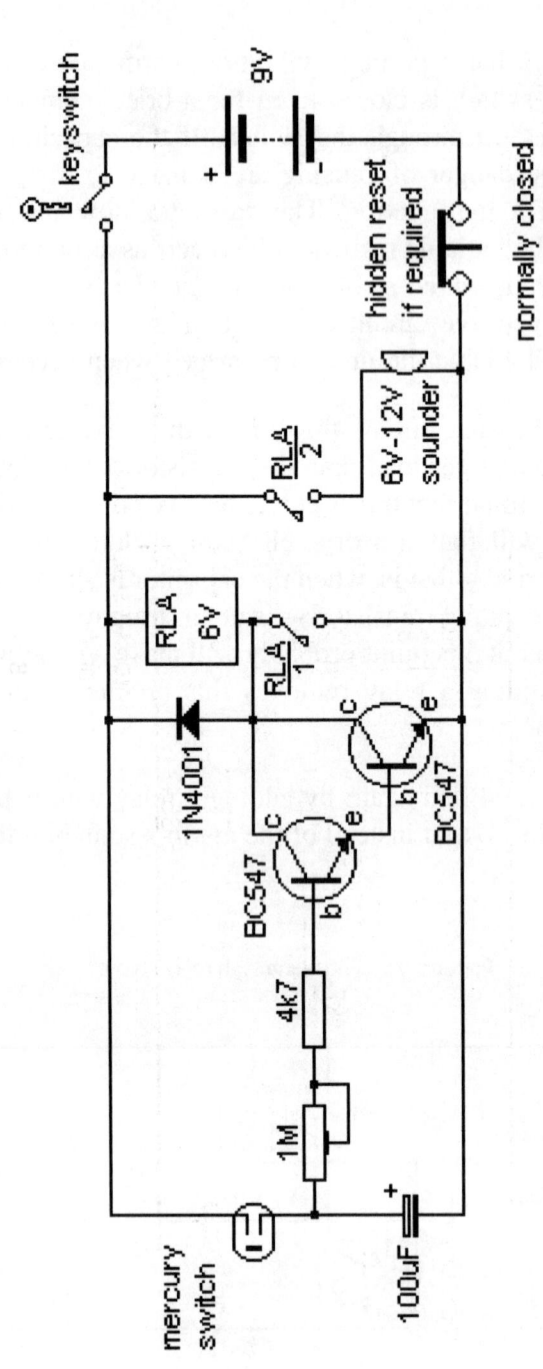

FIGURE 8.73

MISCELLANEOUS DESIGNS 227

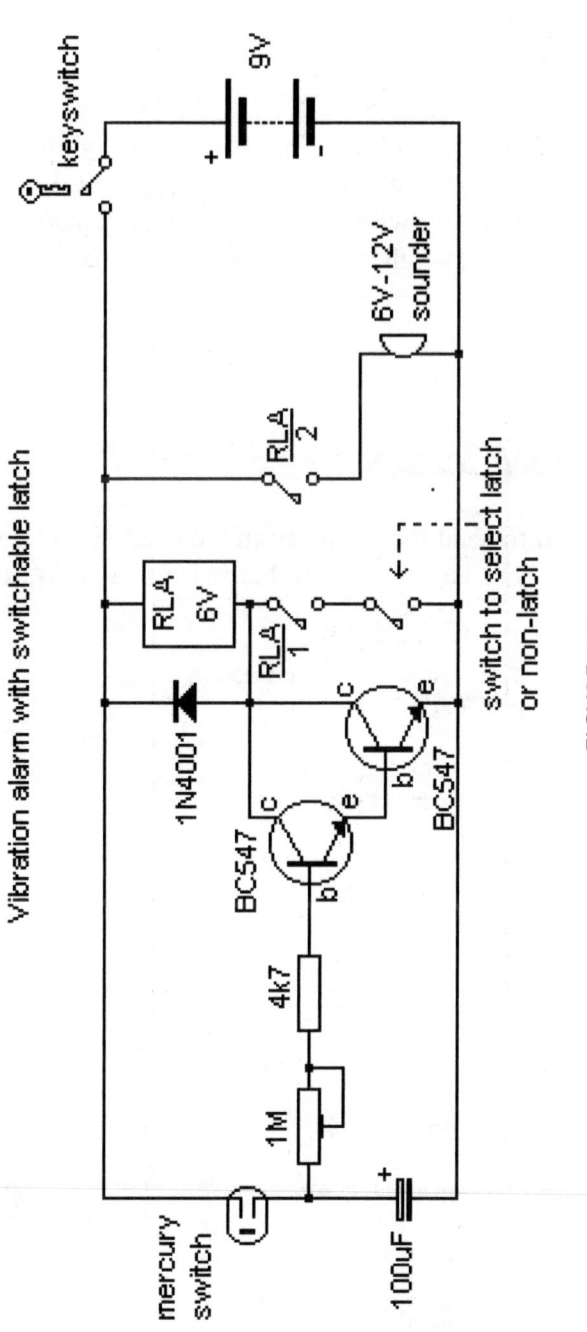

FIGURE 8.74

Vibration alarm with switchable latch

while, it will latch on until reset by the key switch. In this design, the timing capacitor is not required.

The fourth circuit, Figure 8.74, has a modification providing the additional capability of selecting either a latching or nonlatching unit. With an additional switch, or a wire link in series with the latching relay contact, the circuit will be latching, whereas if the switch is opened or the wire link removed, latching will be inhibited.

VIDEO TRANSMISSION

Sometimes you need to send the signal from a closed-circuit TV camera to a monitoring area without the means of hardwiring; that is, some form of

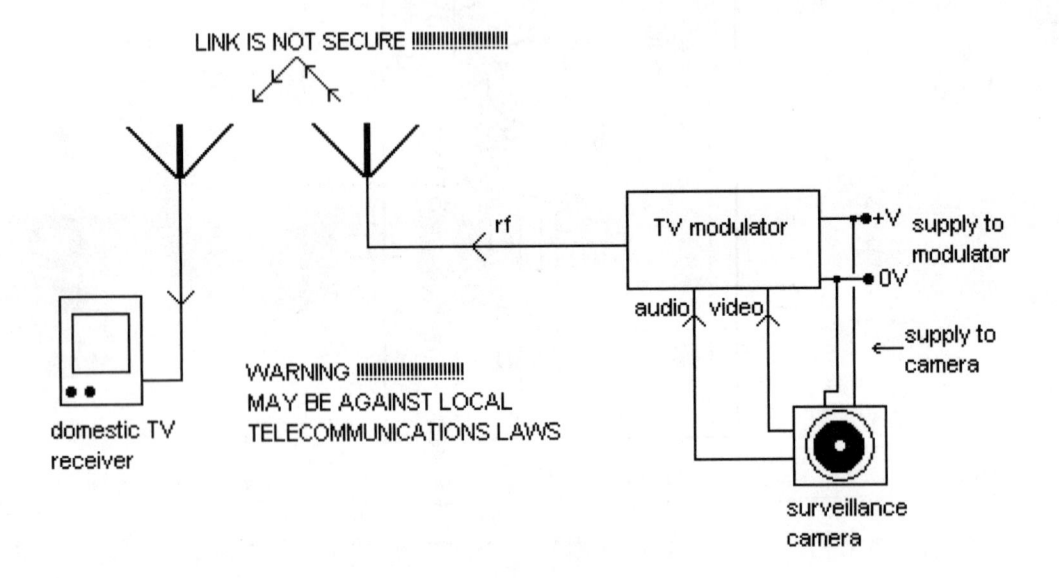

FIGURE 8.75

transmitter is required. Perhaps the most simple method to provide short-range transmission is to utilize a standard television modulator, possibly one salvaged from an old video game or an old computer. The method of connection is shown in Figure 8.75, along with an associated power supply connection schematic. Note that by definition, the system is not secure, so the possibility of someone tuning in to the signal, although remote, is real! Also note that most countries have legislation prohibiting use of video transmission, no matter how weak the signal. The range of this system will depend on the type of modulator used and may be increased only by having high-gain aerial systems on both transmitter and receiver.

Should you have no scrap modulator at hand, the circuit shown in Figure 8.76 can be utilized. Because it is not easy to build a UHF transmitter, this device will operate at around 65MHz, at the bottom end of the television VHF band. With the output of the transmitter's being somewhat unclean, you may be able

Low power video transmitter

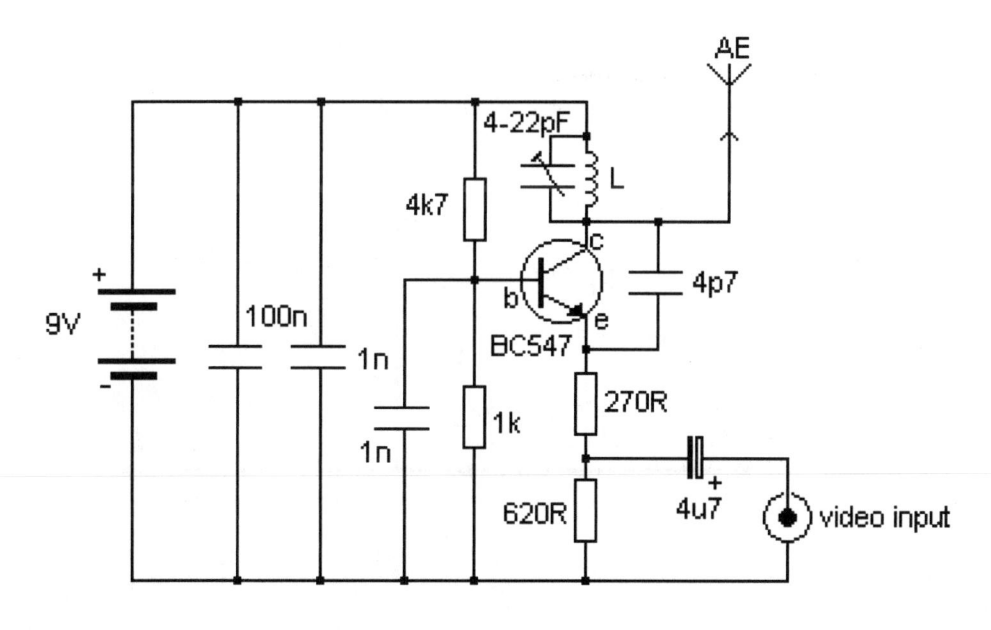

FIGURE 8.76

to view the harmonic transmissions on a UHF receiver, albeit at a much decreased distance.

The tuning coil "L" is made from winding around a dozen turns of 22-gauge enamelled wire on a 3/8-inch former, such as a piece of plastic tubing, old marker pen case, or the like. The aerial can be a few feet of insulated wire, unless you require a more elaborate design. The circuit is a standard free-running, or VFO, design, with modulation being applied to the emitter of the transistor to provide amplitude modulation. Tuning of the transmitter is done by altering either of the L or C components in the tuned circuit. This means either stretching the coil to increase the operating frequency or compressing it to decrease the frequency. Alternatively, if a 4-22pF trimmer capacitor is used as shown in the diagram, this may be adjusted to alter the frequency. The resistor in the emitter circuit may be altered to improve the modulation of the transmission, because undermodulation will produce a weak, low-contrast type of picture, whereas overmodulation will cause the picture to tear up and become unusable.

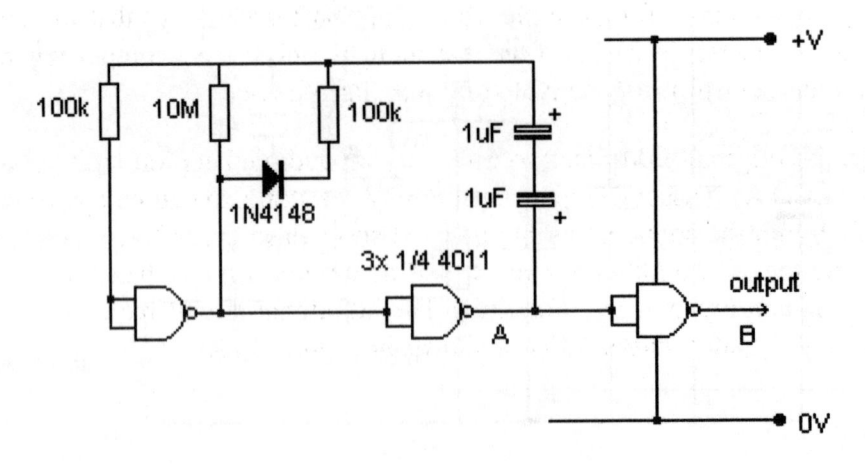

FIGURE 8.77

FIGURE 8.77a

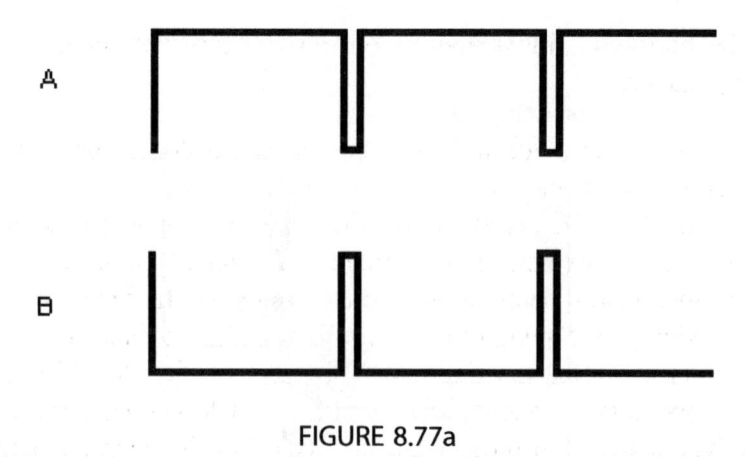

WARNING BEEP CIRCUIT

A reliable, low–component count beep circuit that generates a short beep every five seconds is shown in Figure 8.77. Although two polarized electrolytic capacitors are used in this design, you may use a single unpolarized capacitor if one is available. The output waveforms at points A and B are shown in Figure 8.77a, indicating the unequal mark-to-space ratio generated by this circuit. The circuit readily lends itself to indicators that warn of blown fuses, low batteries, and the like, by using a control voltage on one of the gate input connections—for example, the output buffer gate—so that the circuit maybe connected in such a way that there is only a beep output when the additional control pin is activated by the alarm device.

Although originally designed to give only an audible beep on fault, an additional LED with associated current-limiting resistor also can be connected in parallel with the sounder, so that a visual short flash every five seconds also is permitted. The additional LED would be a boon if more than one beeping circuit is employed at any one time! The additional circuit requirements for driving a sounder and/or LED is shown in Figure 8.77b.

Warning Circuit with audio-visual output

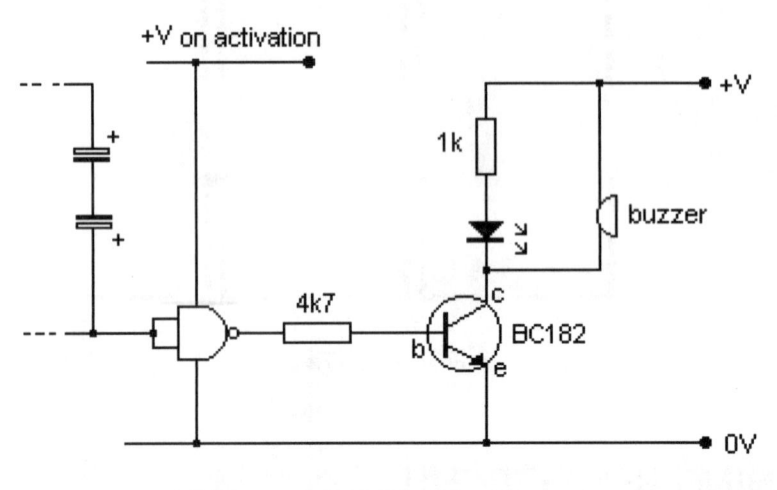

FIGURE 8.77b

WATER/FLUID-LEVEL DETECTORS

Safety Notice. Do not under any circumstances use any noncommercial device to measure the level of inflammable liquids !!!

There are many uses for detecting water and other fluids, such as for warning of basement or container flooding, rain warnings, and so on. Figure 8.78 shows a fluid alarm that is based on the 555 timer IC mentioned in Chapter 3, Timing Circuits. Pin 4 is the reset pin of the device, which, if left disconnected or connected to ground, will inhibit the device. If the fluid level rises, thereby connecting the pin to the positive rail, the circuit produces an audible warning. Note that as in all designs of this nature, these designs depend on the fluid being at least slightly conductive. Water from the ground usually contains sufficient ions to give the water some kind of relatively good conductivity, whereas, say, water produced from a still would be unmeasurable.

A circuit similar to the preceding one is shown in Figure 8.79. This design also uses a 555 timer IC but is somewhat more conventional. This audible

MISCELLANEOUS DESIGNS | 233

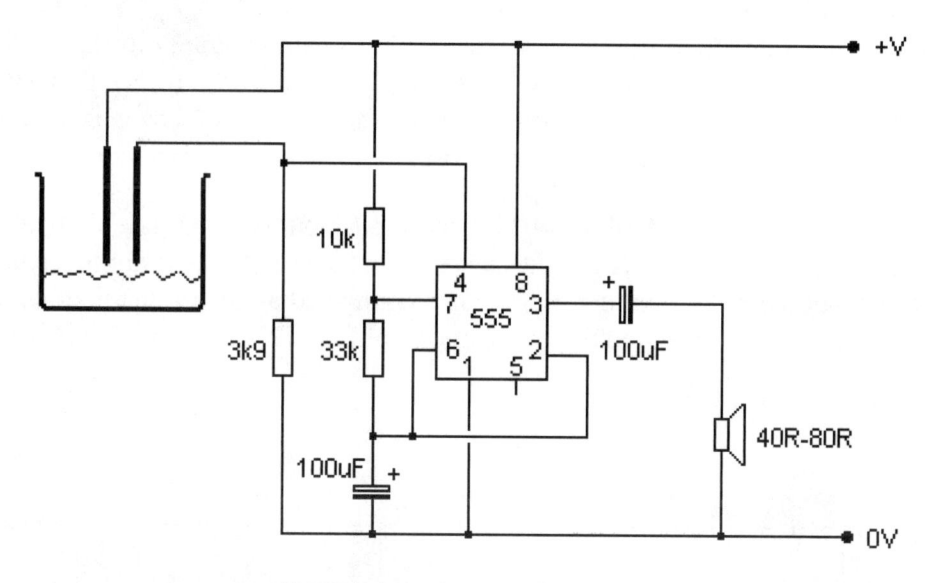

FIGURE 8.78 Water / fluid alarm

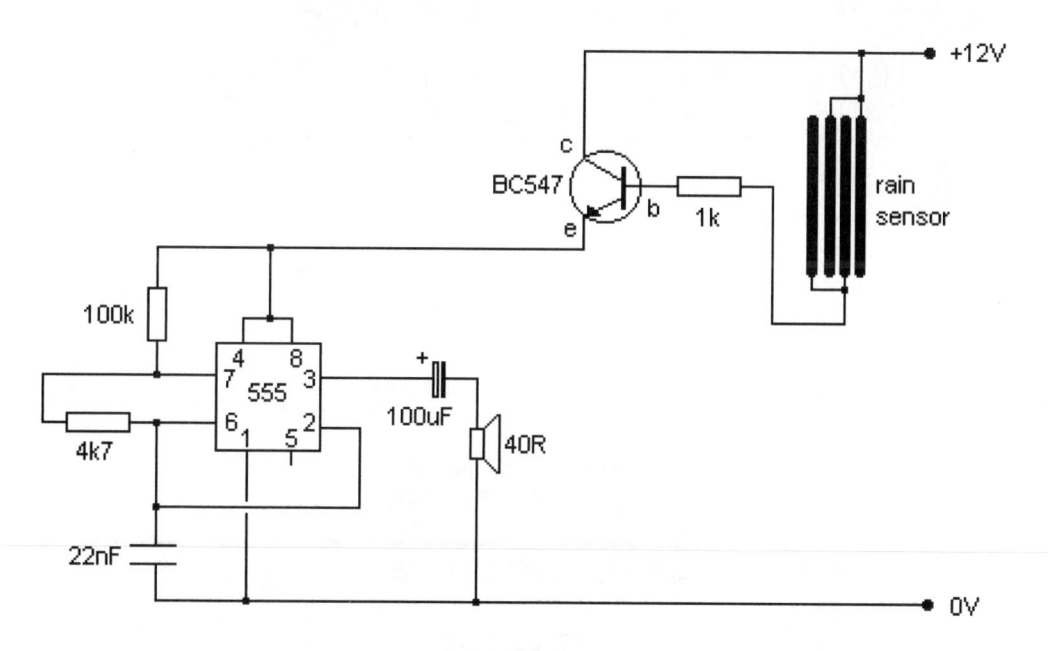

FIGURE 8.79 Rain/fluid Sensor using 555 IC

warning device will be triggered by turning the positive supply to the IC on and off by using the transistor as a switch. When the base of the transistor becomes positive, as the rain, for example, shorts out the sensor rails, the transistor switches on and so powers up the multivibrator circuit.

A circuit based on the CMOS 4011 quad dual-input NAND gate device is shown in Figure 8.80. This circuit uses the first two gates as a noninverting buffer input, with the two remaining gates connected as an oscillator so as to provide an audible output to the transistor amplifier. This sensitive device can

Damp detector using 4011

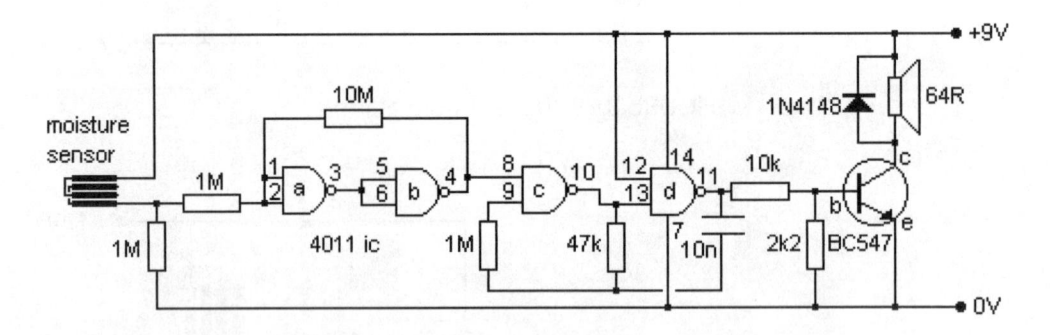

FIGURE 8.80

method of fabricating moisture / rain sensor from copper stripboard

FIGURE 8.80a

be used as a rain sensor, but it also may be used as a meter for checking dampness in buildings walls. A method of fabricating a water/rain detector from stripboard is shown in Figure 8.80a, and the pinout of the 4011 device is shown in Figure 8.81.

It may be necessary to be able to remotely monitor the different levels of a fluid rather than just the two states of "yes, there is" or "no, there isn't." To this end, Figure 8.82 shows a method of continuously remotely monitoring the fluid level by using a series of detectors. The last detector may be wired to an alarm to give a warning that the level is dangerously high. If the conductivity of the fluid is too small, the single transistor may be replaced by an amplifier, as shown in Figure 8.83.

Top View of 4011 ic

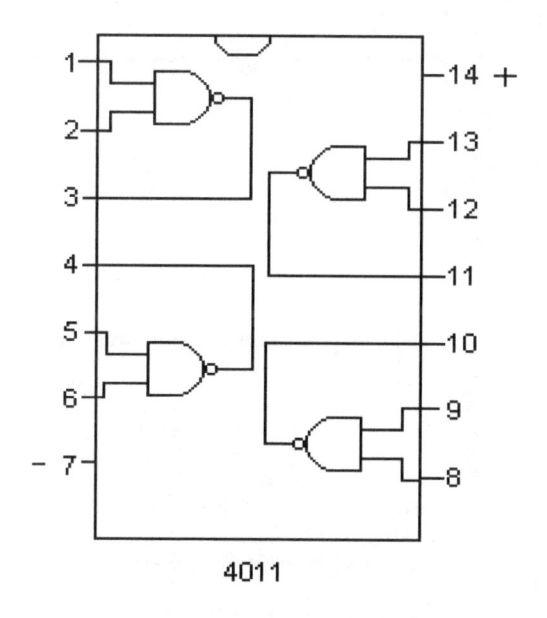

4011

FIGURE 8.81

APPLIED SECURITY DEVICES AND CIRCUITS

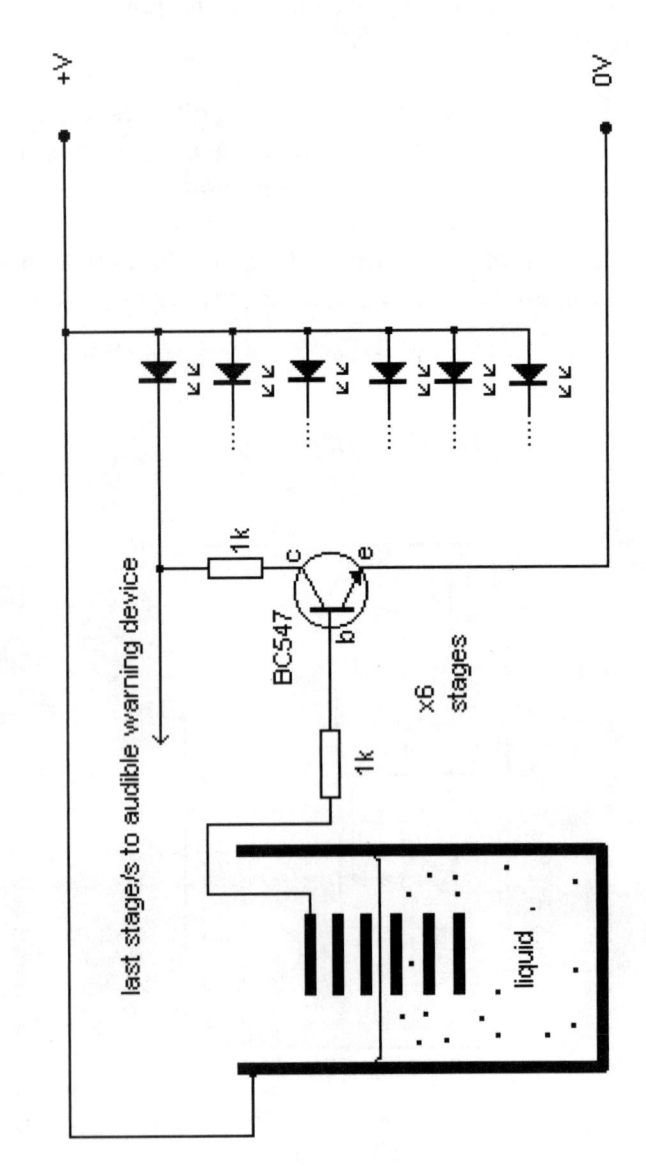

FIGURE 8.82 Water level remote measurement

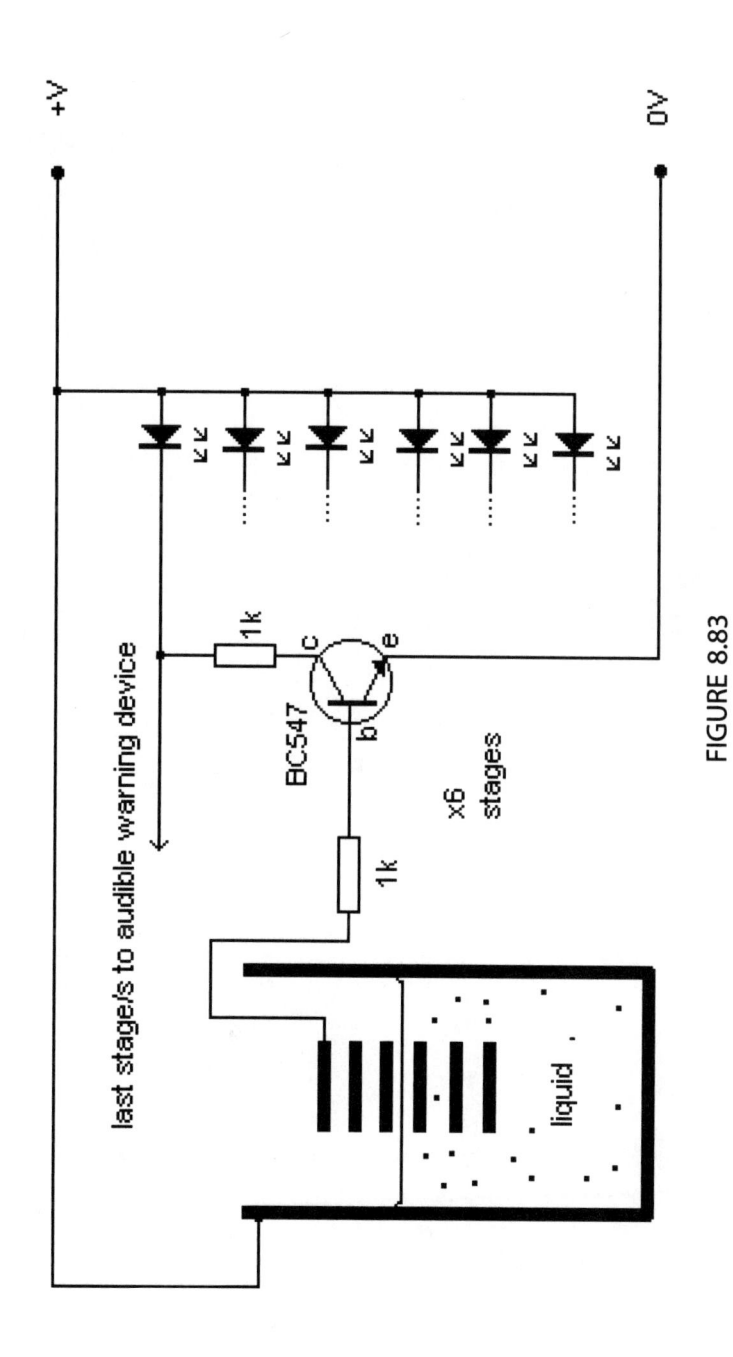

FIGURE 8.83

Water level remote measurement

last stage/s to audible warning device

BC547

x6 stages

liquid

Index

A

AC 105
adio-frequency interference 140
alarm, fishbite 151
alarm fob 61
alarm, touch-activated 213
alternating current 105
ammeter 21
amplifier, high-gain 164
amplifier transistor 202
amplifier, transistor 179
AND gate 162
anti–tamper switch 112
attenuator 179
audible output 197
audio oscillator 84
automatic-dialing system 13
automobile ignition coil 105

B

backup battery 117
beat-frequency oscillator 86
BFO 86
BNC 146
bridge rectifier 105

buffer 21
button, panic-alarm 2

C

camera 19
camera, CCD 223
capacitor, decoupling 135
capacitor, electrolytic 68, 140, 189, 200
capacitor, smoothing 117
capacitor, trimmer 81, 135
carbon microphone 15
CCD 19
CCD camera 223
charge-coupled device 19
circuit, detector 134
circuit, inverter 104
circuit, latch 213
circuit, latching 23, 120, 197
circuit, light-timing 116
circuit, multivibrator 154, 218
circuit, self-latching 132
circuit, time-out 197
circuit, VFO 136
clocking signal 183
closed-circuit TV system 2

crystal oscillator 85
crystal-controlled oscillator 85
current-limiting resistor
    168, 171, 197, 220
current-sensing alarm 71

D

daisy chain 31, 36
damp detector 166
Darlington Pair 68, 72, 180,
    193, 197, 225
DC 105
decoupling capacitor 135
detector circuit 134
detector coil 175
digital volt meter 60
diode 67
diode, rectification 117
diode, silicon 85
diode, zener 67, 97, 187
double-switching contact 212
dual-input NAND gate 234
dummy device 143
DVM 60
dynamic microphone 15

E

electret microphone 15
electrode, grounding 102
electrolytic capacitor 68, 140,
    187, 200
electromechanical device 21
electronic sounder 20
electronic thermostat 94
end-of-line resistor 32
EOL 32

F

fishbite alarm 151
four-wire system 36
full-wave bridge rectifier 105
fuse, quick-blow 61

G

gas discharge device 105
GDD 107
glass-mounted sensor 16
grounding electrode 102

H

Hall-effect switch 14
Hall-effect transistor 14
Hartley oscillator 104
high-gain amplifier 164
hum sensor switch 155
hyperbright 144

I

in-line socket 167
indicator relay 67
infrared 223
infrared receiver 161
infrared transmission 161
input device 23

input/sensing device 12
intercom 3
intermediate frequency 179
inverted input 91
inverter 105
inverter circuit 104

J

jack plug 120
jack socket 120

K

key fob 61
key sentinel 162
key switch 168

L

latch circuit 213
latching circuit 23, 120, 197
latching contact 168
latching relay contact 168
LC circuit 85
LC oscillator 177
LCD 21
LDR 14, 25, 65, 94, 187, 193
LED 21, 58, 61, 74, 97, 117, 133, 165, 197, 220
LED cluster 223
light-dependent resistor 14, 94, 187
light-sensing switch 187

light-timing circuit 116
lightning detector 207
lock, solenoid 123
logic gate 21
loop alarm system 166

M

magnetic reed switch 26
magnetic sensing 17
magnetic switch 12
microphone 15
microphone, carbon 15
microphone, dynamic 15
microphone, electret 15
microphone, piezoelectric crystal 15
microswitch 13
modulator 229
multivibrator circuit 154, 218

N

NAND gate 132, 183, 234
NAND logic gate 202
negative temperature coefficient 94
negative temperature coefficient thermistor 78
noninverting input 91, 97
NOR gate 87
npn transistor 218
NTC 94

O

one-conductor system 100
one-cycle circuit 212
op-amp 91, 92, 95, 120
op-amp device 74
op-amp IC 78
op-amp switch 210
opto device 12
oscillator, audio 84
oscillator, beat-frequency 86
oscillator, crystal 85
oscillator, crystal-controlled 85
oscillator, Hartley 104
oscillator, LC 177
oscillator, reference 179
oscillator, search 179
oscillator, variable-frequency 81
output device 23, 26
output transistor 74

P

panic switch 13
panic-alarm button 2
passive infrared detector 15
pcb 63
peephole 2
photovoltaic cell 15
piezo-resistive principle 17
piezoelectric crystal microphone 15
PIR 15
PIR sensor 112
plunger switch 13

PNP transistor 137
polarity-sensitive 133
power supply unit 145
preamplifier 158
pressure mat 14, 26
pressure sensor 17
pressure switch 197
printed circuit board 63
probe 19
proximity sensor 16
PSU 145

Q

quick-blow fuse 61

R

radio-controlled activator 193
radio-frequency interference 95
radio-frequency sensing 18
radio-frequency spectrum 84
RC network 87
rectification diode 117
rectifier, bridge 105
reed relay 21
reed switch 21, 26
reference oscillator 179
relay 21
relay, reed 21
relay, self-latching 160
relay, solid-state 22
relay, transistor-switched 149
remote sensor 13

resistor, current-limiting
    168, 171, 197, 220
resistor, light-dependant 94
resistor, light-dependent 14, 187
resistor, variable 68, 95, 104
RF 18
RF connector 146
RF decoupling capacitor 135
RF energy 135
RFI 95, 140

S

search head 175
search oscillator 179
security alarm system 2, 4
self-latching circuit 132
self-latching relay 160
sensor, glass-mounted 16
sensor, PIR 112
sensor, pressure 17
sensor, remote 13
sensor, sound 15
sensor switch 126, 127
sensor, vibration 12, 16
side tone 84
signal processor 23
silicon diode 85
sink current 87
smoothing capacitor 117
solenoid 21, 22
solenoid lock 123
solid-state relay 22
sound sensor 15
sound-activated switch 198

spring-biased switch 140
static electricity detector 207
static heat source 15
superheterodyne system 175
switch, anti–tamper 112
switch, Hall-effect 14
switch, hum sensor 155
switch, light-sensing 187
switch, magnetic 12
switch, op-amp 210
switch, panic 13
switch, plunger 13
switch, pressure 197
switch, reed 21, 26
switch, sensor 126, 127
switch, sound-activated 198
switch, spring-biased 140
switch, tamper 12, 109,
    111, 140, 151
switch, temperature-activated 210
switch, tilt 14, 224
switch, transistor 126
switch, trembler 13, 68
switch, voice-operated 133
switched transistor 128

T

tamper switch 12, 109,
    111, 140, 151
temperature-activated switch 210
temperature-sensitive resistor 16
thermistor 16, 25, 210
thermostat 78, 80
thermostat, electronic 94

thermostatically controlled heating switch 210
thyristor 105
tilt switch 14, 224
time-out circuit 197
touch-activated alarm 213
tracking transmitter 81
transistor amplifier 179
transistor, amplifier 202
transistor, Hall-effect 14
transistor, npn 218
transistor, output 74
transistor, PNP 137
transistor switch 126
transistor, switched 128
transistor-switched relay 149
trembler switch 13, 68
trimmer capacitor 81, 135
two-conductor system 103
two-wire system 34

U

ultrasonic detector 16
ultrasonic system 146

V

variable resistor 68, 95, 104
variable-frequency oscillator 81
vehicle tracking device 81
very high frequency 84
VFO 81, 84
VFO circuit 136
VHF 84
vibration sensor 16
vibration sensor 12
visual indicator 21
voice-operated switch 133
volt meter, digital 60
voltage regulator 117
voltmeter 21
VOX 133
VOX recorder 133

W

window foil 12
wire loop 12
wire loop system 167

Z

zener diode 67, 97, 187

EXPLORING LANS FOR THE SMALL BUSINESS & HOME OFFICE

Author: LOUIS COLUMBUS
ISBN: 0790612291 • **SAMS#:** 61229
Pages: 304 • **Category:** Computer Technology
Case qty: TBD • **Binding:** Paperback
Price: $39.95 US/$63.95CAN

About the book: Part of Sams Connectivity Series, *Exploring LANs for the Small Business and Home Office* covers everything from the fundamentals of small business and home-based LANs to choosing appropriate cabling systems. Columbus puts his knowledge of computer systems to work, helping entrepreneurs set up a system to fit their needs.

PROMPT® Pointers: Includes small business and home-office Local Area Network examples. Covers cabling issues. Discusses options for specific situations. Includes TCP/IP (Transmission Control Protocol/Internet Protocol) coverage. Coverage of protocols and layering.

Related Titles: *Administrator's Guide to E-Commerce*, by Louis Columbus, ISBN 0790611872. *Administrator's Guide to Servers*, by Louis Columbus, ISBN

Author Information: Louis Columbus has over 15 years of experience working for computer-related companies. He has published 10 books related to computers and has published numerous articles in magazines such as *Desktop Engineering, Selling NT Solutions*, and *Windows NT Solutions*. Louis resides in Orange, Calif.

To order today or locate your nearest PROMPT® Publications distributor at 1-800-428-7267 or www.samswebsite.com

Prices subject to change.

EXPLORING MICROSOFT OFFICE XP

Authors: JOHN BREEDEN & MICHAEL CHEEK
ISBN: 079061233X • **SAMS#:** 61233
Pages: 336 • **Category:** Computer Technology
Case qty: TBD • **Binding:** Paperback
Price: $29.95 US/$47.95CAN
About the book: Breeden and Cheek provide an insight into the newest product from Microsoft — Office XP. Office XP is the replacement for Microsoft Office, designed to take users into the 21st century. Breeden and Cheek provide tips and tricks for the experienced office user, to help them find maximum value in this new software.

ELECTRONICS FOR THE ELECTRICIAN

Author: NEWTON C. BRAGA
ISBN: 0790612186 • **SAMS#:** 61218
Pages: 320 • **Category:** Electrical Technology
Case qty: 32 • **Binding:** Paperback
Price: $34.95 US/$55.95CAN
About the book: Author Newton Braga takes an innovative approach to helping the electrician advance his or her career. Electronics have become more and more common in the world of the electrician, and this book will help the electrician become more comfortable and proficient at tackling the new tasks required.

To order today or locate your nearest PROMPT® Publications distributor at 1-800-428-7267 or www.samswebsite.com

Prices subject to change.

ADMINISTRATOR'S GUIDE TO SERVERS

Author: LOUIS COLUMBUS
ISBN: 0790612305 • **SAMS#:** 61230
Pages: 304 • **Category:** Computer Technology
Case qty: TBD • **Binding:** Paperback
Price: $39.95 US/$63.95CAN

About the book: Part of Sams Connectivity Series, *Administrator's Guide to Servers* piggybacks on the success of Columbus' best-selling title *Administrator's Guide to E-Commerce*. Columbus takes a global approach to servers while providing the detail needed to utilize the correct application for your Internet setting.

PROMPT® Pointers: Compares approaches to server development. Discusses administration and management. Balance of hands-on guidance and technical information.

Related Titles: *Administrator's Guide to E-Commerce*, by Louis Columbus, ISBN 0790611872. *Exploring LANs for the Small Business and Home Office*, by Louis Columbus, ISBN 0790612291. *Computer Networking for the Small Business and Home Office*, by John Ross, ISBN 0790612216.

Author Information: Louis Columbus has over 15 years of experience working for computer related companies. He has published 10 books related to computers and has published numerous articles in magazines such as *Desktop Engineering, Selling NT Solutions*, and *Windows NT Solutions*. Louis resides in Orange, Calif.

To order today or locate your nearest PROMPT® Publications distributor at 1-800-428-7267 or www.samswebsite.com

Prices subject to change.

SEMICONDUCTOR CROSS REFERENCE BOOK, 5/E

Author: SAMS TECHNICAL PUBLISHING
ISBN: 0790611392 • **SAMS#:** 61139
Pages: 876 • **Category:** Professional Reference
Case qty: 14 • **Binding:** Paperback
Price: $39.95 US/$63.95CAN

About the book: The perfect companion for anyone involved in electronics! Sams has compiled years of information to help you make the most of your stock of semiconductors. Both paper and CD-ROM versions of this tool contain an additional 128,000 parts listings over the previous editions.

ON CD-ROM, 2E
ISBN: 0790612313 • **SAMS#:** 61231 • **Price:** $39.95 US/$63.95CAN

COMPUTER NETWORKS FOR THE SMALL BUSINESS & HOME OFFICE

Author: JOHN A. ROSS
ISBN: 0790612216 • **SAMS#:** 61221
Pages: 304 • **Category:** Computer Technology
Binding: Paperback • **Price:** $39.95 US/$63.95CAN
About the book: Small businesses, home offices, and satellite offices with unique networks of 2 or more PCs can be a challenge for any technician. This book provides information so that technicians can install, maintain and service computer networks typically used in a small business setting. Schematics, graphics and photographs will aid the "everyday" text in outlining how computer network technology operates, the differences between various network solutions, hardware applications, and more.

To order today or locate your nearest PROMPT® Publications distributor at 1-800-428-7267 or www.samswebsite.com

Prices subject to change.

ADMINISTRATOR'S GUIDE TO DATAWAREHOUSING

Author: AMITESH SINHA
ISBN: 0790612496 • **SAMS#:** 61249
Pages: 304 • **Category:** Computer Technology
Case qty: TBD • **Binding:** Paperback
Price: $39.95 US/$63.95CAN

About the book: Datawarehousing is the manipulation of the data collected by your business. This manipulation of data provides your company with the information it needs in a timely manner, in the form it desires. This complex and emerging technology is fully addressed in this book. Author Amitesh Sinha explains datawarehousing in full detail, covering everything from set-up to operation to the definition of terms.

PROMPT® Pointers: Covers On-Line Analytical Processing issues. Addresses set-up of datawarehousing systems. Is designed for the experienced IT administrator.

Related Titles: *Designing Serial SANS*, ISBN 0790612461, *How the PC Hardware Works*, ISBN 079061250X.

Author Information: Amitesh Sinha has a Masters in Business Administration and over 10 years of experience in the field of Information Technology. Sinha is currently the Director of Projects with GlobalCynex Inc. based in Virginia and has written numerous articles for computer publications.

To order today or locate your nearest PROMPT® Publications distributor at 1-800-428-7267 or www.samswebsite.com

Prices subject to change.

APPLIED ROBOTICS II

Author: EDWIN WISE
ISBN: 0790612224 • **SAMS#:** 61222
Pages: 304 • **Category:** Projects
Case qty: TBD • **Binding:** Paperback
Price: $29.95 US/$47.95CAN

About the book: Edwin Wise builds upon his best-seller, *Applied Robotics* with this book targeted at more advanced hobbyists with development of a larger, more robust, and very practical mobile robot platform. Building on the foundation set in his first text, *Applied Robotics II* has projects to create a larger robot platform suitable for use in the home or outdoors, advanced sensor projects and a great exploration of A1 and control software.

Prompt Pointers: Picks up where *Applied Robotics* left off. Offers an advanced set of projects related to this very hot subject area.

Related Titles: *Applied Robotics*, ISBN 0790611848. *Animatronics*, ISBN 079061294.

Author Information: Edwin Wise is a professional software engineer with twenty years of experience. He currently works in the field of Computer Aided Manufacturing (CAM). His experience includes work on both computer games and educational software. Building robots has been a dream and passion for Edwin for years now. His current project is "Boris," a giant killer robot that can be viewed at http://www.simreal.com/Boris.

To order today or locate your nearest PROMPT® Publications distributor at 1-800-428-7267 or www.samswebsite.com

Prices subject to change.

GUIDE TO CABLING AND COMMUNICATION WIRING

Author: LOUIS COLUMBUS
ISBN: 0790612038 • **SAMS#:** 61203
Pages: 320 • **Category:** Communications
Case qty: TBD • **Binding:** Paperback
Price: $39.95 US/$63.95CAN

About the book: Part of Sams Connectivity Series, *Guide to Cabing and Communication Wiring* takes the reader through all the necessary information for wiring networks and offices for optimal performance. Columbus goes into LANs (Local Area Networks), WANs (Wide Area Networks), wiring standards and planning and design issues to make this an irreplaceble text.

PROMPT® Pointers:
Features planning and design discussion for network and telecommunications applications. Explores data transmission media. Covers Packet Framed-based data transmission.

Related Titles: *Administrator's Guide to E-Commerce*, by Louis Columbus, ISBN 0790611872. *Exploring LANs for the Small Business and Home Office*, by Louis Columbus, ISBN 0790612291. *Computer Networking for the Small Business and Home Office*, by John Ross, ISBN 0790612216.

Author Information: Louis Columbus has over 15 years of experience working for computer-related companies. He has published 10 books related to computers and has published numerous articles in magazines such as *Desktop Engineering, Selling NT Solutions*, and *Windows NT Solutions*. Louis resides in Orange, Calif.

To order today or locate your nearest PROMPT® Publications distributor at 1-800-428-7267 or www.samswebsite.com

Prices subject to change.

HOW THE PC HARDWARE WORKS

Author: MICHAEL GRAVES
ISBN: 079061250X • **SAMS#:** 61250
Pages: 800 • **Category:** Computer Technology
Case qty: TBD • **Binding:** Paperback
Price: $39.95 US/$63.95CAN

About the book: As the technology surrounding our desktop PCs continues to evolve at a rapid pace, the opportunity to understand, repair and upgrade your PC is attractive. In an era where the PC you bought last year is now "out of date", your opportunity to bring your PC up-to-date rests in this informative text. Renouned author Michael Graves addresses this subject in a one-on-one manner, explaining each category of computer hardware in a complete, concise manner.

Prompt Pointers: Designed to bring a beginner up to a professional level of hardware expertise. Includes new SCSI III implementations, new video standards, and previews of upcoming technologies.

Related Titles: *Exploring Office XP*, ISBN 079061233X, *Designing Serial SANS*, ISBN 0790612461, *Administrator's Guide to Datawarehousing*, ISBN 0790612496.

Author Information: Michael Graves is a Senior Hardware Technician and Network Engineer for Panurgy of Vermont. Graves has taught computer hardware courses on the college level at Champlain College in Burlington, Vermont and The Essex Technical Center in Essex Junction, Vermont. While this is his first full-length book under his own name, his contributions have been included in other works and his technical writing has been the source of several of the more readable user's guides and manuals for different products.

To order today or locate your nearest PROMPT® Publications distributor at 1-800-428-7267 or www.samswebsite.com

Prices subject to change.

AUTOMOTIVE AUDIO SYSTEMS

Author: HOMER L. DAVIDSON
ISBN: 0790612356 • **SAMS#:** 61235
Pages: 320 • **Category:** Automotive
Case qty: TBD • **Binding:** Paperback
Price: $39.95 US/$63.95CAN

About the book: High-powered car audio systems are very popular with today's under-30 generation. These top-end systems are merely a component within the vehicle's audio system, much as your stereo receiver is a component of your home audio and theater system. Little has been written about the troubleshooting and repair of these very expensive automotive audio systems. Homer Davidson takes his decades of experience as an electronics repair technician and demonstrates the ins-and-outs of these very high-tech components.

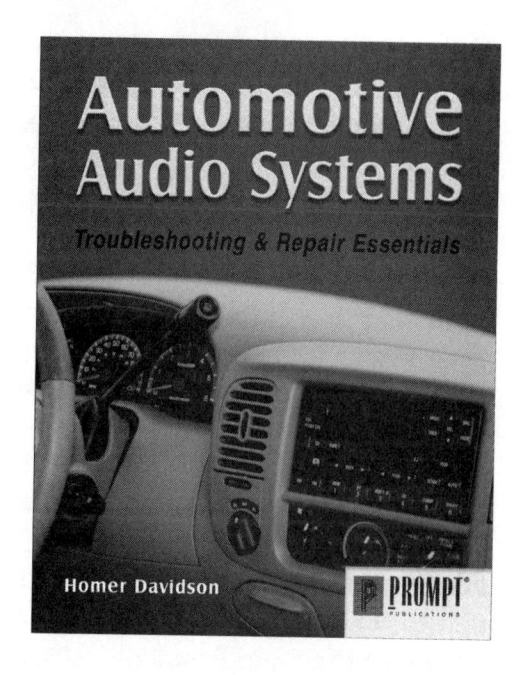

Prompt Pointers: Coverage includes repair of CD, Cassette, Antique car radios and more. All of today's high-end components are covered. Designed for anyone with electronics repair experience.

Related Titles: *Automotive Electrical Systems*, ISBN 0790611422. *Digital Audio Dictionary*, ISBN 0790612011. *Modern Electronics Soldering Techniques*, ISBN 0790611996.

Author Information: Homer L. Davidson worked as an electrician and small appliance technician before entering World War II teaching Radar while in the service. After the war, he owned and operated his own radio and TV repair shop for 38 years. He is the author of more than 43 books for TAB/McGraw-Hill and Prompt Publications. His first magazine article was printed in *Radio Craft* in 1940. Since that time, Davidson has had more than 1000 articles printed in 48 different magazines. He currently is TV Servicing Consultant *for Electronic Servicing & Technology* and Contributing Editor for *Electronic Handbook*.

To order today or locate your nearest PROMPT® Publications distributor at 1-800-428-7267 or www.samswebsite.com

Prices subject to change.

DESIGNING SERIAL SANS

Author: WILLIAM DAVID SCHWADERER
ISBN: 0790612461 ● **SAMS#:** 61246
Pages: 320 ● **Category:** Computer Technology
Case qty: TBD ● **Binding:** Paperback
Price: $39.95 US/$63.95CAN

About the book: The use of Serial SANS is an increasingly popular and efficient way to store data in a medium to large corporation setting. Serial SANS effectively stores your company's data away from the traditional server, allowing your valuable server resources to be used for running applications.

Prompt Pointers: Covers Device Specialization Considerations. Explains Media Signals, Data Encoding and Protocols. Discusses SAN hardware building blocks.

Related Titles: *Administrator's Guide to Datawarehousing*, ISBN 0790612496, *How the PC Hardware Works*, ISBN 079061250X.

Author Information: W. David Schwaderer has extensive complex computer system experience and was involved in the creationof two Silicon Valley start-up companies. Schwaderer has a diverse background in connectivity products, personal computer software, and voice DSP based systems. Schwaderer currently resides in Saratoga, Calif.

ADMINISTRATOR'S GUIDE TO THE EXTRANET/INTRANET

Author: CONRAD PERSSON
ISBN: 0790612410 • **SAMS#:** 61241
Pages: 304 • **Category:** Computer Technology
Case qty: TBD • **Binding:** Paperback
Price: $34.95 US/$55.95CAN

About the book: We are all familiar with the Internet, but few of us have occasion to utilize an Intranet or Extranet application. Both have vast applications related to inner-company communication, customer service, and vendor relations. Both are built similarl to Internet sites, and have many of the same features, issues, and problems. Intranet and Extranet applications are generally under-utilized, even though they provide the opportunity for both communication and financial benefits.

Prompt Pointers: Designed for the Systems Administrator or advanced webmaster. Outlines Intranet/Extranet issues, problems, and opportunities. Discusses hardware and software needs.

Related Titles: *Administrators Guide to E-Commerce*, ISBN 0790611872. *Computer Networking for the Small Business and Home Office*, ISBN 0790612216. *Exploring Microsoft Office XP*, ISBN 079061233X.

Author Information: Conrad Persson is the editor of *ES&T Magazine*, the premier publication for the electronics servicing industry. Conrad has decades of experience related to electronics and computer applications and resides in Shawnee Mission, Kan.

To order today or locate your nearest PROMPT® Publications distributor at 1-800-428-7267 or www.samswebsite.com

Prices subject to change.

BROADBAND EXPOSED

Author: MICHAEL BUSBY
ISBN: 0790612488 • **SAMS#:** 61248
Pages: 352 • **Category:** Communications
Case qty: TBD • **Binding:** Paperback
Price: $39.95 US/$63.95CAN

About the book: As the telecommunications industry goes through deregulation, the lines between telecom and other communication applications have become very blurred. Business mergers and advancing technology have created a need for more and more broadband applications. Author Michael Busby addresses these issues in relation to the telecommunications sector as well as topics pertaining to cable, satellite, RF, microwave and other communication methods. A must read for anyone working with telecommunication technologies.

Prompt Pointers: Includes discussions of LAN, CAD, imaging, wire, cable and more. Addresses networking fundamentals, protocols, and multimedia applications.

Related Titles: *Telecommunication Technologies*, ISBN 0790612259, *Exploring LANS for the Small Business & Home Office*, ISBN 0790612291, *Guide to Cabling & Communication Wiring*, ISBN 0790612038.

Author Information: Michael Busby is president and CEO of Mikal Enterprises, a global telecommunications design and consulting company. Busby has over 30 years telecommunications experience as field service engineer, systems engineer, R&D engineer, engineering manager, product manager, and VP engineering.

To order today or locate your nearest PROMPT® Publications distributor at 1-800-428-7267 or www.samswebsite.com

Prices subject to change.

GUIDE TO DEDICATED MICROPROCESSOR FUNDAMENTALS

Author: BOB ROSE
ISBN: 0790612402 • **SAMS#:** 61240
Pages: 320 • **Category:** Electronics Technology • **Binding:** Paperback
Price: $34.95 US/$55.95CAN

About the book: Dedicated Microprocessors are found in almost every consumer electronic device on the market today. Although not a repairable component, the dedicated microprocessor holds the clues to many successful repairs. This text discusses structure, function, communications, operating fundamentals, failures, requirements, peripheral problems and more.

Prompt Pointers: Dedicated microprocessors are found in most electronic devices. Technicians of all levels need to understand dedicated microprocessors to effectively repair devices. Little has been written on dedicated microprocessors related to repairs.

Related Titles: *Manufacturer to Manufacturer Part Number Cross Reference with CD-ROM*, ISBN 0790612321. *Semiconductor Cross Reference Guide 5E*, ISBN 0790611392. *DSP Filters*, ISBN 0790612046.

Author Information: Bob Rose has spent his career in the electronics servicing industry. An expert at the troubleshooting and repair of TVs and VCRs, Rose has and continues to service all the major brands of electronic devices. Bob Rose holds more than 40 training certificates and is the author of more than 30 articles and two books. As a member of the National Electronics Service Dealers Association, Bob stays abreast of the latest trends in technology from his home in Medina, Tenn.

To order today or locate your nearest PROMPT® Publications distributor at 1-800-428-7267 or www.samswebsite.com

Prices subject to change.

HOME THEATER SYSTEMS

Author: BOB GOODMAN
ISBN: 0790612372 • **SAMS#:** 61237
Pages: 304 • **Category:** Video Technology
Case qty: TBD • **Binding:** Paperback
Price: $39.95 US/$63.95CAN

About the book: In days past, you had a TV, radio, and maybe a turntable in your "living room." Today, the evolution of electronics has brought us the Home Theater System, combining projection TVs, high-powered audio receivers, multiple CD players, DVD systems, surround-sound and more. This plethora of components is rarely purchased from a single manufacturer, making installation and maintenance a complicated task at best. Bob Goodman applies his electronics experience to this topic and provides a guidebook to home theater systems, including information on systems, components, troubleshooting, and maintenance.

Prompt Pointers: Home theater systems are the future of home audio/video systems. A buyer's guide is included. Great detail is provided regarding component choices.

Related Titles: *Digital Audio Dictionary*, ISBN 0790612011. *DVD Player Fundamentals*, ISBN 0790611945. *Guide to Satellite TV Technology*, ISBN 0790611767.

Author Information: Bob Goodman, CET, has devoted much of his career to developing and writing about more effective, efficient ways to troubleshoot electronics equipment. An author of more than 62 technical books and 150 technical articles, Goodman spends his time as a consultant and lecturer in Western Arkansas.

To order today or locate your nearest PROMPT® Publications distributor at 1-800-428-7267 or www.samswebsite.com

Prices subject to change.

BASIC SOLID STATE ELECTRONICS

Author: VAN VALKENBURG
ISBN: 0790610426 • **SAMS#:** 61042
Pages: 944 • **Category:** Electronics Technology
Case qty: 12 • **Binding:** Paperback
Price: $29.95 US/$47.95CAN
About the book: Considered to be one of the best books on solid-state electronics on the market, this revised edition provides the reader with a progressive understanding of the elements that form various electronic systems. Electronic fundamentals covered in the illustrated, easy-to-understand text include semiconductors, power supplies, audio and video amplifiers, transmitters, receivers, and more.

CMOS SOURCEBOOK

Author: NEWTON BRAGA
ISBN: 0790612348 • **SAMS#:** 61234
Pages: 304 • **Category:** Electronics Technology
Case qty: TBD • **Binding:** Paperback
Price: $39.95 US/$63.95CAN
About the book: CMOS (Complementary Metal Oxide Semiconductors) are an essential part of almost every electronics component and are not typically understood. Braga takes the concepts from the legendary CMOS Cookbook from Don Lancaster (originally published by Sams/Macmillan) and brings them into the 21st Century with this new and different look at CMOS IC technology.

To order today or locate your nearest PROMPT® Publications distributor at 1-800-428-7267 or www.samswebsite.com

Prices subject to change.

ADMINISTRATOR'S GUIDE TO E-COMMERCE

Author: LOUIS COLUMBUS
ISBN: 0790611872 ● **SAMS#:** 61187
Pages: 416 ● **Category:** Business Technology
Case qty: 28 ● **Binding:** Paperback
Price: $34.95 US/$55.95CAN
About the book: Unlike previous electronic commerce books which stress theory, the Administrator's Guide to E-Commerce is a hands-on guide to creating and managing Web sites using the Microsoft BackOffice product suite. This book will explore the role of networking technologies to industry growth, issues of privacy and security, and most importantly, guidance in taking an existing Web server and creating an electronic storefront.

ANIMATRONICS: GUIDE TO HOLIDAY DISPLAYS

Author: EDWIN WISE
ISBN: 0790612194 ● **SAMS#:** 61219
Pages: 304 ● **Category:** Projects
Case qty: 32 ● **Binding:** Paperback
Price: $29.95 US/$47.95CAN
About the book: Author Edwin Wise takes the reader inside his world of robotics in an innovative guide to designing, developing, and building animated displays centered around the holidays of Halloween and Christmas.

To order today or locate your nearest PROMPT® Publications distributor at 1-800-428-7267 or www.samswebsite.com

Prices subject to change.

GUIDE TO ELECTRONIC SURVEILLANCE DEVICES & CIRCUITS

Author: CARL BERGQUIST
ISBN: 0790612453 • **SAMS#:** 61245
Pages: 304 • **Category:** Video Technology
Case qty: TBD • **Binding:** Paperback
Price: $32.95 US/$52.50CAN

About the book: One of the hottest topics of 2001, Surveillance, is covered in dept in this text from Carl Bergquist. The accessability of electronic components, Internet access and affordable parts coupled with the increased fears and concerns of the general public has created a boom in the surveillance industry. Bergquist covers the building of surveillance systems including video surveillance, wireless systems, computer network systems, and audio systems. Also discussed are issues related to this sensitive topic, uses, legal considerations, counter-surveillance and much more.

Prompt Pointers: One of the hottest electronic topics of 2001! Includes 6 do-it-yourself projects. Designed for anyone with electronics knowledge.

Author Information: Carl Bergquist followed a successful career as a photojournalist for AP, UPI, *The New York Times*, *Newsweek*, and other publications by turning his efforts toward a lifelong hobby of electronics. Besides articles in *Popular Electronics* and *Electronics Now*, Carl has authored numerous books for Prompt Publications.

To order today or locate your nearest PROMPT® Publications distributor at 1-800-428-7267 or www.samswebsite.com

Prices subject to change.

BASIC ELECTRICITY

Author: VAN VALKENBURG
ISBN: 0790610418 ● **SAMS#:** 61041
Pages: 736 ● **Category:** Electrical Technology
Case qty: 16 ● **Binding:** Paperback
Price: $29.95 US/$47.95CAN
About the book: Considered to be one of the best electricity books on the market, the authors have provided a clear understanding of how electricity is produced, measured, controlled and used. A minimum of mathematics is used for direct explanations of primary cells, magnetism, Ohm's Law, capacitance, transformers, DC generators, and AC motors. Other essential topics covered include conductance, current flow, electromagnetism and meters.

Best Seller!

BASIC ELECTRICITY AND DC CIRCUITS

Author: CHARLES DALE
ISBN: 0790610728 ● **SAMS#:** 61072
Pages: 928 ● **Category:** Electronics Technology
Case qty: 10 ● **Binding:** Paperback
Price: $39.95 US/$63.95CAN
About the book: No matter what their background, readers can learn the basic concepts that have enabled mankind to harness and control electricity. Chapters are arranged to allow readers to progress at their own pace, with concepts and terms being introduced as needed for comprehension.

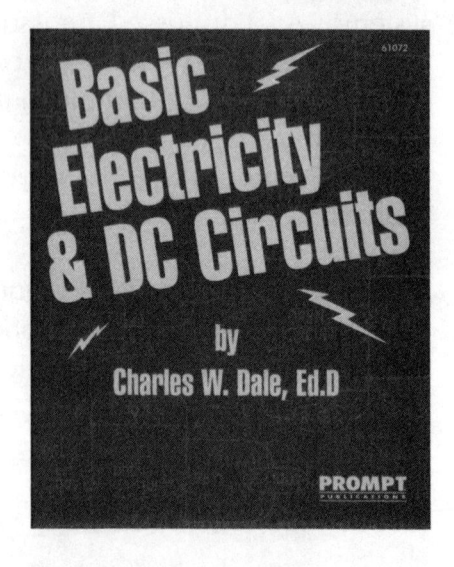

To order today or locate your nearest PROMPT® Publications distributor at 1-800-428-7267 or www.samswebsite.com

Prices subject to change.

AUTOMOTIVE ELECTRICAL SYSTEMS

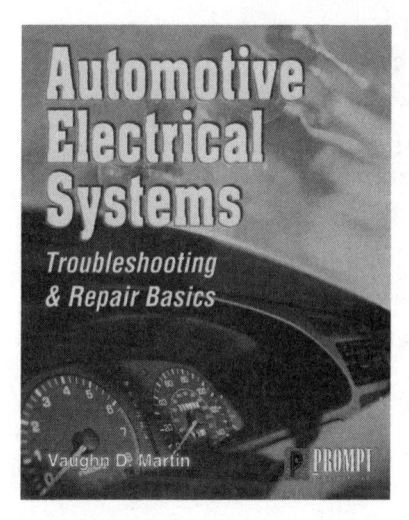

Author: VAUGHN D. MARTIN
ISBN: 0790611422 • **SAMS#:** 61142
Pages: 272 • **Category:** Automotive
Case qty: 36 • **Binding:** Paperback
Price: $29.95 US/$47.95CAN

About the book: Chilton manuals assume that you already know enough to fully understand and fix the problem you are having. Automotive Electrical Systems fills that knowledge and experience gap, allowing you to better understand your car's electrical systems and computers well enough to buy appropriate replacement parts and then install them. Automotive Electrical Systems walks you through the reading of repair schematics, which are vastly different on automobiles than on traditional consumer electronics devices.

DSP FILTER COOKBOOK

Author: JOHN LANE, ET AL
ISBN: 0790612046 • **SAMS#:** 61204
Pages: 344 • **Category:** Electronics Technology
Case qty: 26 • **Binding:** Paperback
Price: $39.95 US/$63.95CAN

About the book: Digital filters and real-time processing of digital signals have traditionally been beyond the reach of most, due partially to hardware cost as well as complexity of design. In recent years, low-cost digital signal processor (DSP) development boards have put this within reach. This book will break down this design complexity barrier by means of simplified tutorials, step-by-step instructions, along with a collection of audio projects.

To order today or locate your nearest PROMPT® Publications distributor at 1-800-428-7267 or www.samswebsite.com

Prices subject to change.

GUIDE TO DIGITAL CAMERAS

Author: MICHAEL MURIE
ISBN: 0790611759 • **SAMS#:** 61175
Pages: 536 • **Category:** Video Technology
Case qty: 18 • **Binding:** Paperback
Price: $39.95 US/$63.95CAN
About the book: The Complete Guide to Digital Cameras will appeal to anyone who has recently purchased or is considering an investment in a digital camera. Together the book and CD-ROM will answer questions you have about digital cameras, enable you to make intelligent buying decisions, and help you use your camera to its full potential. No camera purchase is complete without this informative guide.

HOME AUTOMATION II -LITETOUCH SYSTEMS

Author: JAMES VAN LAARHOVEN
ISBN: 0790612267 • **SAMS#:** 61226
Pages: 336 • **Category:** Projects
Case qty: 32 • **Binding:** Paperback
Price: $34.95 US/$55.95CAN
About the book: James Van Laarhoven explores the very comprehensive home automation system from LiteTouch Systems. This book will aid in the installation, maintenance and programming of the LiteTouch 2000. Includes lighting, audio, installation, blueprint reading, video and more.

To order today or locate your nearest PROMPT® Publications distributor at 1-800-428-7267 or www.samswebsite.com

Prices subject to change.